Use R!

Series Editors
Robert Gentleman   Kurt Hornik   Giovanni Parmigiani

For a complete list of titles published in this series, go to www.springer.com/series/6991

Use R!

Andrew P. Robinson • Jeff D. Hamann

Forest Analytics with R

An Introduction

 Springer

Andrew P. Robinson
Dept. Mathematics and Statistics
University of Melbourne
Parkville 3010 VIC
Australia
A.Robinson@ms.unimelb.edu.au

Jeff D. Hamann
Forest Informatics, Inc.
PO Box 1421
97339-1421 Corvallis Oregon
USA
jeff.hamann@forestinformatics.com

Series Editors:
Robert Gentleman
Program in Computational Biology
Division of Public Health Sciences
Fred Hutchinson Cancer Research Center
1100 Fairview Avenue, N. M2-B876
Seattle, Washington 98109
USA

Kurt Hornik
Department of Statistik and Mathematik
Wirtschaftsuniversität Wien Augasse 2-6
A-1090 Wien
Austria

Giovanni Parmigiani
The Sidney Kimmel Comprehensive
Cancer Center at Johns Hopkins University
550 North Broadway
Baltimore, MD 21205-2011
USA

ISBN 978-1-4419-7761-8   e-ISBN 978-1-4419-7762-5
DOI 10.1007/978-1-4419-7762-5
Springer New York Dordrecht Heidelberg London

Springer is part of Springer Science+Business Media (www.springer.com)

This book is dedicated to Grace, Felix, and M'Liss,
(and Henry and Yohan)
with grateful thanks for inspiration and patience.

Preface

R is an open-source and free software environment for statistical computing and graphics. R compiles and runs on a wide variety of UNIX platforms (e.g., GNU/Linux and FreeBSD), Windows, and Mac OSX. Since the late 1990s, R has been developed by hundreds of contributors and new capabilities are added each month. The software is gaining popularity because: 1) it is platform independent, 2) it is free, and 3) the source code is freely available and can be inspected to determine exactly what R is doing.

Our objectives for this book are to 1) demonstrate the use of R as a solid platform upon which forestry analysts can develop repeatable and clearly documented methods; 2) provide guidance in the broad area of data handling and analysis for forest and natural resources analytics; and 3) to use R to solve problems we face each day as forest data analysts.

This book is intended for two broad audiences: students, researchers, and software people who commonly handle forestry data; and forestry practitioners who need to develop actionable solutions to common operational, tactical, and strategic problems. We often mention better and more complete treatments of specific subject material for further reference (e.g., forest sampling, spatial statistics, or operations research).

We hope that this book will serve as a field manual for practicing forest analysts, managers, and researchers. We hope that it will be dog-eared, defaced, coffee/tea-stained, and sticky-noted to near destruction. We hope the reader will engage in the exercises, scrutinize its contents, forgive our weaknesses, possibly and carefully absorb suggestions, and constructively criticize.

Acknowledgments

This book would not have been possible without the patient and generous assistance of many people. We first thank all the authors of the literature we cite, who were willing to publish their data as part of their research.

These data are often our only link between repeatable research and anecdotal opinion.

We thank Valerie LeMay and Timothy Gregoire for their kind contribution of tree measurement data and for their encouragement and leadership in the field. We thank Boris Zeide for his generous contribution of the von Guttenberg data. We thank Don Wallace and Bruce Alber for supplying an interesting dataset to demonstrate the data management, plotting, and file functions in Chapter 2. We thank the Oregon State University College of Forestry Research Forests web site for posting a publicly available forest inventory for Chapters 2 and 4. Without those data, many of the examples and topics in this book would have to have been performed using simulated data and frankly would have been much less interesting. We thank Martin Ritchie for providing data, funding, and snippets of code once lost and found again during the development of the rconifers package, used in Chapter 8. We thank David Hann for, years ago, providing an original copy of the manuscript that we used to generate the shared library example (`chambers-1980.so`) in Chapter 8 (Chambers, 1980).

We have received considerable constructive criticism via the review process, only some of which we can source. We especially thank John Kershaw for generous and detailed comments on Chapter 3, Jeff Gove for his support and useful commentary on Chapter 5, and David Ratkowsky and Graham Hepworth for their thoughtful and thought-provoking comments on Chapters 6 and 7, respectively. Numerous other useful comments were made by anonymous reviewers. The collection of review comments improved the book immeasurably.

We thank R Core, the R community, and all the package authors and maintainers we have come to rely upon. Specifically, we thank the following people, in no particular order: David B. Dahl (xtable); Lopaka Lee (R-GLPK); Andrew Makhorin (GLPK); Roger Bivand (maptools); Deepayan Sarkar (lattice); Hadley Wickham (ggplot2); Brian Ripley (MASS, class, boot); Jose Pinheiro and Douglas Bates (nlme); Frank Harrell (Hmisc); Alvaro Novo and Joe Schafer (norm); Greg Warnes (gmodels et al.); Reinhard Furrer, Douglas Nychka, and Stephen Sain (fields); Thomas Lumley (survey); and Nicholas J. Lewin-Koh and Roger Bivand (maptools).

We thank John Kimmel, our managing editor at Springer, for showing incredible patience and Hal Henglein, our copy editor, for keeping us consistent.

AR wishes to thank Mark Burgman for providing space to finish this book within a packed ACERA calendar, and for his substantial support and guidance. AR also wishes to thank Geoff Wood, Brian Turner, Alan Ek, and Albert Stage for their kindness and intellectual support along the way.

JH wishes to thank Martin Ritchie, David Marshall, Kevin Boston, and John Sessions for their support along the way.

Finally, we wish to thank our wives, children, and friends for cheerful perseverance and support in the face of a task that seemed at times like a little slice of Sisyphus.

Melbourne and Corvallis   *Andrew Robinson*
August 16, 2010   *Jeff D. Hamann*

Contents

Part III Allometry and Fitting Models

Part I
Introduction and Data Management

Chapter 1
Introduction

Forestry has offered a fertile environment for data analysts to operate in since forest measurements began. Forestry datasets are typically voluminous, hierarchical, messy, multi-faceted, and expensive. The challenges that forest managers face are complex, and the costs of poor decisions can be high. On the other hand, often, but not always, decisions are made over long time frames, and there is time for considered data analysis. Forestry data analysts are fortunate to work in a space in which the data and the resources are usually sufficient to do something that is more useful than doing nothing.

Forestry has a wide range of datasets and questions in which statistics, econometrics, and applied mathematics tools can all play a constructive role. The challenge for the data analyst is to find the best match between the data, the question, and the tools. The match depends on the context. A model or dataset that perfectly suits one application may be quite inappropriate for another. The same model or dataset may be just the best that can be done at the time, and reluctantly accepted. The analyst must be pragmatic.

This need for pragmatism must cut across received dogma from statistics and other fields. For example, in Chapters 6 and 7 we spend considerable time scrutinizing graphical diagnostics to learn more about the intersections between particular models and the data to which we have fit them, as encapsulated in the model assumptions. The importance of the fidelity of the match between the assumptions and the diagnostics depends entirely on context. We can learn more about the data and can possibly improve the model, but at some point the analysis finishes and the action begins. To be clear: sometimes you have to make the decision in front of you using the data and the model that you have rather than the data and the model that you wish you had. Go ahead.

This book develops and demonstrates solutions to common forestry data-handling and analysis challenges. We draw upon solutions from applied statistics, forest biometrics, and operations research. Most of these solutions have already been suggested and applied in forestry literature; our goal is to survey them, and demonstrate an approach to resolving them using R (Ihaka and

Gentleman, 1996). R is an open-source, GNU-licensed statistical programming language that has interpreters for several computing platforms, notably Unix, Windows, and Mac OSX, in 32 bit and 64 bit versions (R Development Core Team, 2010). R is free, flexible, and powerful, and rapidly making inroads into various aspects of statistical endeavor, including those relevant to forestry.

Forestry datasets are usually collected and managed according to local or organizational customs. Sometimes these customs are documented. Datasets may be censored, analyses chaotic, vocabularies inchoate, and processes more often breached than observed. The data analyst must respond flexibly and creatively, document processes, and leave an unambiguous analytical trail. Analyses should be scripted, and scripts should be documented, time-stamped, and carefully archived. It is in these contingencies that R can shine most brightly. As we will see in Chapter 2, R provides a suite of data-reading and data-handling tools that can be combined to handle pretty much any data-preparation challenge.

A cornucopia of analytical strategies is coupled to this plethora of data. R encompasses a considerable range of statistical and mathematical tools by itself, and with its extension packages, written by the R community, its reach widens further. The count of such packages, at the time of writing, is in the multiple thousands. The quality of these packages, in terms of the match between what they say and what they do, is not guaranteed. Some packages are used and scrutinized daily by hundreds of individuals. Some packages are developed in passing, contributed to the community, and rarely dusted off. Although there has been discussion of the development of a systematic review facility or the tracking of usage statistics, these have not yet been developed. The analyst must consider the source carefully.

The challenge of real-world data analysis can be divided into substantial parts. In an ideal situation, these challenges are met sequentially. More often, iteration, if not outright back-tracking, is necessary. The parts are:

1. Abstraction: the art of translation of a practical question into a model. This process is interactive and relies at least as much upon communication skills and sensitivity as it does statistical ability.
2. Data collection: from the field, the filing cabinet, the hard drive, or the Internet, find the data that are most likely to satisfy the model as it pertains to the practical question, within the time and effort that are dictated by the available budget.
3. Modeling: find the best match between the model and the data that are available, in the context of the question that the model represents.
4. Conclusion: stop modeling. Identify the model flaws rigorously and move on.
5. Communication: translate the results and the model flaws back to the answer to the practical question and the caveats with which it must be interpreted.

Regardless of the platform used for analysis, the development of carefully documented scripts will inevitably save anguish and time, sooner or later.

1.1 This Book

The objectives of this book are to 1) demonstrate the use of R as a solid platform upon which forestry analysts can develop repeatable and literate programming methods; 2) provide guidance in the broad area of data handling and analysis for forest and natural resources analytics; and 3) to use R to solve problems we face each day as forest data analysts.

This book is intended for two broad audiences: 1) students, researchers, and software people who commonly handle forestry data; and 2) forestry practitioners who need to develop actionable solutions to common operational, tactical, and strategic problems.

At times, our citations may appear excessive, and at other times lacking. We shamelessly mix units and nomenclature. The focus of this book is to present R as an analytical tool for manipulating data, performing analysis, and generating useful outputs. We mention, where appropriate, better and more complete treatments of the subject material, and assume the reader either has access to these standard texts or understands the details behind the subject matter.

In each chapter, we present one or more analysis problems and at least one solution. Our solutions are not necessarily the most efficient for the problem at hand. We encourage the reader to examine the problem in a variety of ways and develop alternative solutions. In each chapter, we also present a number of questions that are important for many analysis tasks and attempt to provide good templates for providing objective, unbiased, and useful answers.

1.1.1 Topics Covered in This Book

Four major subject areas are covered in this text. Part I includes this introduction and a chapter presenting the fundamental processes of data ingestion, data manipulation, and performance of basic procedures to examine forest, forestry, and forestry-related data.

In Part II, we present the processing of sample surveys, and imputation methods. In Chapter 3, we cover the analysis of forest sample surveys, interval estimation methods, single-level sampling, hierarchical sampling, and samples using auxiliary information. In Chapter 4, we cover mapping, imputation, and prediction of spatial data using nearest-neighbor, expectation maximization, and kriging methods to generate a complete dataset for a forest landscape when we begin with incomplete or missing data.

In Part III, we present more sophisticated tools for examining allometric relationships, fitting dimensional distributions, and linear and non-linear hierarchical models. In Chapter 5, we fit diameter distributions using parametric and non-parametric representations. In Chapter 6, we fit linear and non-linear regression models. In Chapter 7, we present methods for model construction using maximum likelihood, linear mixed-effects models, and non-linear mixed-effects models.

In Part IV, we present techniques for simulation and optimization in forest environments. In Chapter 8, we shift focus to methods to generate simulations for disparate forest models using C and FORTRAN code. In Chapter 9, we present and solve a well-known linear forest estate planning and optimization problem to determine the wood flow for a strict area harvest schedule (Leuschner, 1990).

1.1.2 Conventions Used in This Book

This book contains numerous equations, examples of source code, computer output, and citations. The format for the computer inputs and outputs in this book is as follows: 1) names of files, objects, and functions used within the text are in `courier` font; 2) package names are in plain text; 3) R code that you type at the command line is in *slanted courier* font; and 4) output that comes out of R is in `courier` font. For example, a simple R session would look like (startup message omitted):

```
> a <- rnorm(1000)
> b <- mean(a)
> b

[1] -0.01262165
```

where a and b are objects of type vector and scalar; rnorm and mean are functions that return objects, respectively.

1.1.3 The Production of the Book

This book was produced by the authors using LaTeX and the Sweave function in R, and was compiled under a variety of platforms. At times, both authors have used the Windows, FreeBSD, and Mac OSX operating systems, including both 32 bit and 64 bit versions. While we make no guarantees that these scripts will work on other platforms, we encourage readers to report their results.

1.2 Software

As well as demonstrating forest analytics, another aim of this book is to enable the reader to quickly get started in using these methods via R. The R software is compiled for numerous platforms and can be freely obtained from the Comprehensive R Archive Network at http://cran-r.project.org.

1.2.1 Communicating with R

R is, at its most basic level, a command-line language, meaning that we interact with R by typing sequences of commands at its prompt. The default prompt, which means that R is ready to receive an instruction, looks like this:

```
>
```

Sometimes input to R will be longer than one line. In these circumstances, R will change the prompt to let you know that it expects a continuation of previous input. By default, this is the plus sign:

```
+
```

A common trap that starting R users fall into is to fail to notice which prompt is being provided. If you get a + when you expect a >, then R is still waiting for input to complete. This completion could require the balancing of parentheses, or quotes, for example. If you get lost, then cancel out and start again. Canceling out is a platform-specific operation; for example, in Unix, hitting Ctrl-C does the trick, whereas in Windows there is a big red button to click using the mouse.

We type commands at the prompt, and if R understands them, then it carries them out and returns the result. For example,

```
> 1 + 2
```

```
[1] 3
```

If the command does not explicitly require feedback, then R will not provide it. Thus, if we were to wish to save the result of the preceding arithmetic as an object, called a.out, then we might type

```
> a.out <- 1 + 2
```

and no output is returned. <- is the assignment operator. This is how we tell R to create (or recreate) the object named a.out. Note that R is case-sensitive.

One convenient way to gather sequences of commands together in R is to write them as a function. For example, the following function evaluates the preceding arithmetic expression and provides a cheery message:

```
> hi.there <- function() {
+    a.out <- 1 + 2
+    cat("Hello World!\n")
+    return(a.out)
+ }
```

Some points need elaboration: the call to function creates a new object that is stored in random-access computer memory (RAM). The object is called hi.there, and it contains the function that we have defined. We call the function as follows.

```
> hi.there()
```

```
Hello World!
[1] 3
```

If we wanted to write a more general function, for example one that would add 1 to an arbitrary number, we would include an argument to pass the arbitrary number to the function, as follows.

```
> hi.there <- function(arbitrary.number) {
+      a.out <- 1 + arbitrary.number
+      cat("Hello World!\n")
+      return(a.out)
+ }
> hi.there(pi)
```

```
Hello World!
[1] 4.141593
```

We can feed a sequence of R commands to the command line using the source function. The source function accepts as its first argument the name of a file that contains the R commands to be executed.

The code that we executed above created objects inside R. These objects are stored in RAM in a container referred to as the *workspace*. We can identify the objects in our workspace using the ls function:

```
> ls()
```

```
[1] "a"         "a.out"     "b"         "hi.there"
```

and delete any of them using the rm function:

```
> rm(a.out)
> ls()
```

```
[1] "a"         "b"         "hi.there"
```

Unless we take steps to explicitly save the objects, or the workspace that contains them, they will be lost when the R session finishes. See the help information about the `save` and `save.image` functions for more information. We can delete all the objects in the workspace using

```
> rm(list = ls())
```

This line of code shows us that R will allow us to nest statements seamlessly. Here, the output of the call to `ls` is being used as the `list` argument for the function `rm`.

When we move objects from the workspace to the hard drive or back, as we do in the next chapter, we need to tell R in which directory to look for files or to which directory to save files. If we do not tell R what directory to use, it will use a default directory, called the *working directory*. We can see what the working directory is by using the `getwd` function and set the working directory by using the `setwd` function.

In addition to the command line, R has graphical user interfaces in varying states of sophistication. Adoption of one of these interfaces can simplify the challenge of learning R.

1.2.2 Getting Help

There are four main sources of assistance: the internal help files, the R manuals, the R-help community's archive, and the R-help community itself.

1.2.2.1 Getting Help Locally

While working through this book, it is likely that you will find commands that are used in example code that have been inadequately explained or perhaps ignored completely. When you find such commands, you should read about them using the `help` function, which has `?` as a prefix-style shortcut. We can get help on commands this way; for example

```
> ?mean
> help(mean)
```

This approach is useful so long as we know the name of the command that we wish to use. If we only know some relevant words, then we can use the `help.search` function.

```
> help.search("quartile")
```

The output from this command is long. We have to read through all the descriptions of the functions that correspond to this call until we find the one that seems to be the best match with what we want.

```
...
stats::qqnorm            Quantile-Quantile Plots
stats::quantile          Sample Quantiles
survey::svykm            Estimate survival function.
...
```

Now we can try `help(quantile)`. It also tells us that the `quantile` function is in the stats package. It doesn't tell us that the stats package is already loaded, but it is.

We have found that the best way to learn to use the functions is to try out the examples that usually appear at the end of the help information. For most help files, we can just copy those example commands, paste them to the console, and see what happens. The commands can then be altered to suit our needs. Furthermore, these examples are often miniature data analyses and provide pointers to other useful functions that we can try. Finally,

```
> example(gstat)
```

and

```
> demo(graphics)
```

are useful for running examples from packages.

We can also access the files that are installed with R using a web browser. Again inside R, run the following command:

```
> help.start()
```

This function opens a browser (or a window inside an existing browser) into which help results will be sent and within which you can then point, click, and search for keywords. You may need to set the default browser, depending on your operating system. That would be done by, for example,

```
> options(browser="firefox")
```

The page that is opened in the browser window also provides hyperlinked access to R's manuals, which provide a great deal of very useful information. The manuals are constantly under development, so it is worth checking back regularly. The `options` function allows you to change much of R's default behavior. Use `help` to learn more about it.

1.2.2.2 Getting Help Remotely

There is a thriving community of programmers and users who will be happy to answer carefully worded questions and, in fact, may well have already done so. Questions and answers can be easily found from inside R using the following commands:

```
> RSiteSearch("lme4 p-value", restrict = "Rhelp08")
> RSiteSearch("{logistic regression}") # matches exact phrase
```

If you don't find an answer after a solid search, then you should consider
asking the community by using the R-help email list. There is a posting guide
to help you write questions that are most likely to obtain useful answers —
it is essential reading! Point your browser to `http://www.r-project.org/`
`posting-guide.html` for further guidance.

One point that we emphasize is that a question is much easier to answer
when the motivation can be reproduced. Therefore, if at all possible, include
minimal, commented, executable R code that demonstrates the phenomenon
of interest.

Details on joining the email list group can be found at `https://stat.`
`ethz.ch/mailman/listinfo/r-help`. You may like to consider the digest
option; emails arrive at a rate of up to 100 per day.

1.2.3 Using Scripts

Using R effectively virtually demands that we write scripts. We save the
scripts to a known directory and then either copy and paste them into the R
console or read them in using one of the following commands:

```
> source(file = "C://path/to/scripts/file.R", echo = TRUE)
> source(file = "../scripts/file.R", echo = TRUE)
> source("file.R", echo=TRUE) # If file.R is in
                              # the working directory (q.v.)
```

Note the use of forward slashes to separate the directories. Also, the directory names are case sensitive and are permitted to contain blank spaces.

A key element of good script writing is commentary. In R, the comment
symbol is the # symbol. Everything on a line after a # is ignored. Some
editors will tab comments based on the number of #s used.

Instructions can be delimited by line feeds or semicolons. R is syntacti-
cally aware, so if you insert a return before your parentheses or brackets are
balanced, it will politely wait for the rest of the statement.

Script writing is a very powerful collaborative tool. It's very nice to be able
to send your code and a raw data file to a cooperator and know that they
can just source the code and run your analysis on their machine. Writing
readable, well-commented scripts is a really good habit to get into early; it
makes life much easier in the future. Large projects are vastly simplified by
rigorous script writing.

1.2.4 Extending R

We need to distinguish between the R software application and its packages. When the R application is run, it automatically provides access to a substantial range of functionality. For example, we can compute the mean of a sequence of numbers using the mean function.

```
> mean(c(1, 2, 3))
```

```
[1] 2
```

However, still more functions are available to R within packages that are installed by default but not automatically attached to R's search path. To access these packages, we use the library or require functions. As a modest example, in order to be able to use the boot function for bootstrapping, we must first run

```
> library(boot)
```

We can obtain a list of the packages that are directly available via the library command using the installed.packages function.

In addition to these installed packages, R also provides access to packages that have been written by members of the community. These packages are made available on the CRAN web site, mirrors of which are available in many countries. In order to obtain a list of the packages that are available for download and installation, use the available.packages function. To install one, use the install.packages function.

We have written an R package called FAwR (Forest Analytics with R) that includes the functions and data for this book.

CRAN also offers Task Views, which present collections of packages that support a particular theme. For example, the CRAN Task View for spatial data[1] is an excellent source of options for data and algorithms for spatial data, and the CRAN Task View for Environmetrics[2] is an excellent source for ecological and environmental data. We check the CRAN Task Views regularly and suggest you do, too.

We often use packages that act as an interface to other libraries (for example, GLPK). We refer to the functions as *wrapper* functions, as the goal is to simply pass arguments into the underlying application programming interface (API). For example, in Chapter 9, we use glpk wrapper functions such as lpx_set_mat_row, which stores, or replaces, the contents of the i-th row of the constraint matrix of the specified problem object. As you may find in the R documentation, this function is simply a wrapper for the glpk function, which is written in C.

---

[1] http://cran.r-project.org/web/views/Spatial.html
[2] http://cran.r-project.org/web/views/Environmetrics.html

```
void glp_set_mat_row(glp_prob *lp,
                     int i,
                     int len,
                     const int ind[],
                     const double val[]);
```

where the column indices and numerical values of new row elements must be placed in locations `ind[1]`, . . . , `ind[len]` and `val[1]`, . . . , `val[len]`, respectively. For most if not all of these cases, we refer the reader to the original documents (e.g., Makhorin (2009) for the GLPK API).

1.2.5 Programming Suggestions

Generic programming skills are as useful in coding with R as anywhere else. Wherever possible, examine the results of your code, possibly using summary statistics, to ensure that it has done what you intended. Keep backups of your scripts and datasets. Work with the raw data as it came to you, as much as possible, and execute the necessary cleaning and corrections inside R. This ensures that, when the time comes to share your work, you need only pass on the data as it came to you and one or more R scripts.

We prefer to start with an empty workspace to ensure that no lurking objects can affect code execution. Emptying the workspace is easiest via

```
> rm(list = ls())
```

Of course, these checking steps may later be omitted during automation if that proves clumsy.

For reporting, we have found that the most reliable rounding is done by the `sprintf` function because it retains trailing zeros; for example,

```
> sprintf("%.1f", 3.01)
```

```
[1] "3.0"
```

We find that debugging R code is greatly eased by keeping in mind R's object orientation. We do not go into this aspect of R in any great detail except to mention that if R code is not doing what we expect it to, we often find that it is because the class of the object is not as we expect. We can learn the class of any object by using the `class` function:

```
> class(mean)
```

```
[1] "function"
```

```
> class(c(1, 2, 3))
```

```
[1] "numeric"
```

```
> class(mean(c(1, 2, 3)))
```

```
[1] "numeric"
```

R will often refuse to carry out operations that are inappropriate to the class of the object. This is a feature, not a bug! For example,

```
> not.really.numeric <- c("1", "2", "3")
> class(not.really.numeric)
```

```
[1] "character"
```

```
> mean(not.really.numeric)
```

```
[1] NA
```

Here we can fix the problem by setting the class using one of several approaches.

```
> not.really.numeric <- as.numeric(not.really.numeric)
> class(not.really.numeric)
```

```
[1] "numeric"
```

```
> mean(not.really.numeric)
```

```
[1] 2
```

The second strategy in the debugger's kit is to examine the object. Often the contents or the structure of the object will provide us a hint that our expectations are not being met. Sometimes this realization will lead directly to the solution to our problem. The most useful function in this instance is str, which reports the object's class, its dimensions if appropriate, and a portion of the contents.

```
> str(not.really.numeric)
```

```
 num [1:3] 1 2 3
```

We cannot emphasize enough that the use of str, class, and the related functions dim, head, and tail has led directly to solving problems that could otherwise have taken hours of debugging.

For debugging in more complicated scenarios, for example within functions, we use browser.

1.2.6 Programming Conventions

There are many different ways to do things in R. There are no official conventions on how the language should be used, but the following thoughts may prove useful in communicating with long-time R users.

1. Although the equals sign "=" does work for assignment, it is also used for other things, for example in identifying values for arguments. The arrow "<-" is only used for assignment. We use the arrow for assignment, rather than the equals sign. Others use the equals sign.
2. Spaces are cheap. Use spaces liberally between arguments and between objects and arithmetic operators.
3. Call your objects useful names. Don't call your model `model` or your dataframe `data`.
4. You can terminate your lines with semicolons, but most programmers do not do so.

For example, the following code is hard to read and understand. We don't know what role the constant is playing, and the text is dense.

```
> constant=3.2808399;
> x=x*constant;
```

The following code is easier to read and understand. The identities (and the units) of the variables and the constant are obvious from our naming convention. The equations are spaced so that the distinction between operators and variables is easily seen.

```
> feet_per_meter <- 3.2808399
> heights_m <- heights_ft * feet_per_meter
```

When we return to this code years later, it will be obvious what we did and why we were doing it.

1.2.7 Speaking Other Languages

For processes that take a long time, it may be useful to convert R code into other languages (e.g., C, C++, and FORTRAN) or use other system-accessible tools (glpk, GRASS, or PostgreSQL). We use C or FORTRAN when it seems that our R code will take too long to execute.

For example, in Chapter 8, we build and use a shared library to perform forest simulation for established stands to demonstrate the process. While we use a relatively simple model and a shared library does not appear to be required for our task, we demonstrate all necessary steps required to produce a shared library should the need arise.

Switching and blending languages can make a substantial difference to the overall time taken for a project but has to be balanced against the necessary investment of writing and debugging in another language.

The basic steps that we follow are:

- Write a draft solution in R, and try it out.

- Identify the bottlenecks using the `Rprof` function.
- Write C and FORTRAN functions to replace the computational bottlenecks.
- Compile the functions into a shared library.
- Attach the library to an R session.
- Call the functions when needed.

Note that a constraint in working with FORTRAN is that R can only call FORTRAN subroutines, not functions, and so we must generate wrappers for each of the FORTRAN functions that we wish to call.

Direct access to operating system functions is available via the R `system` function. Judicious use of `system` provides the ability for R to directly oversee the execution of other software on your computer. Personally, we have used `system` at various times to manipulate text files, organize input files, execute simulation runs of an external forest growth simulator (specifically, ORGANON), build our own shared library (`chapters-1980.so`), solve large linear programming problems, and set up the output files for subsequent import and analysis by R and other software.

Other approaches are offered by the inline and Rcpp packages.

1.3 Notes about Data Analysis

The following collection of aphorisms covers ideas that we wish we'd heard earlier, opinions that we wish we'd held earlier, and points for discussion when data analysis seems thornier than it should.

- Find the solution to the simplest possible version of the problem in front of you. Add complexity as you have to.
- Every serious data analysis has multiple phases. Specific problems can be handled in more than one phase of the analysis. Picking the best phase in which to handle each problem is an art.
- The first time you ever submit a statistical report is nerve-wracking. Don't be afraid to start, and don't be afraid to finish (Schabenberger and Pierce, 2002).
- Data can be cheap or expensive. Examine the trade-offs. Select your data wisely.
- Time is often more expensive than computers, hard drives, and memory. Think carefully about the overall investment of time, machines, and creative capital.
- Find the right questions. Traditional statistics provide one perspective. Operations research provides another. Depending on the question, there may not be one best tool for the challenges that you face. Dogmatism will impede your progress.

- Keep aware of the options. Build a satisfactory solution by moving from familiar to unfamiliar territory. Use your background knowledge to construct increasingly more appropriate models. Learn more about the data as the models develop.
- Don't fit a model for your project merely because it is the new thing.
- Fit new models to old projects. Always benefit from your investment of what you've already learned. Use your prior knowledge (about techniques and datasets) as leverage.
- Borrow concepts from other fields to help you examine frameworks for formulating problems (graph theory, sampling, optimization).
- To paraphrase a quote about writing: "There is no good analysis, only good re-analysis."

Chapter 2
Forest Data Management

2.1 Basic Concepts

Proper data management techniques are essential to ensuring flexible and efficient forest resource analysis options. However, little systematic attention has yet been paid to the tools and protocols that are necessary to provide robust and convenient access to data. There exists an overwhelming variety of candidate tools and file formats. This wealth of choices offers the chance to find a solution that best fits an organization's needs. Unfortunately, this same wealth can create problems for interoperability and communication among tools and can create confusion and mistakes and ultimately lead to poor decisions.

For example, many optimization programs use matrix generators to create a file that is then read into another program that solves linear or mixed integer programs. These outputs are then reformatted to conform to some standard organizational report format, which might or might not include graphical displays. The solution from the linear programming software is then transferred to yet another application, which is used for report generation. Those results can then be output to any number of formats or programs such as web browsers, geographic information systems (GIS) applications, or some combination thereof. At each step in transferring and reformatting data, the probability of introducing process errors increases. Performing analysis within a single system can greatly reduce these errors and enhance the analysis process by simplifying data management tasks.

Although it is not necessarily suitable at all scales for all operations, R provides functionality that will allow for the integration of many, and sometimes all, of these different tasks. However, none of these tasks can be accomplished with poor or improper data management techniques. This chapter will focus on those functions within R that read and write files commonly found in forest resource databases, will introduce some basic data-manipulation functions, and will briefly present some of its graphical capabilities. These features, combined with some simple creativity, can create an effective analytical toolbox.

2.2 File Functions

We use the function list.files to report the files that are available in the working directory. To learn what files are available in some other directory, pass the absolute or the relative directory location as an argument. If the relative location is used, it should be located relative to the working directory.

2.2.1 Text Files

The ability to read and write plain text files is critical because text files are very commonly used for storing and communicating data. Most publicly available growth and yield models use plain text files for input and output, as do many landscape-level database systems and cruise compilers. Furthermore, text files can be straightforwardly managed in version control software (CVS and Subversion) to allow users to track all the changes in data over the lifetime of a project. Text files can be easily transferred through email and FTP programs, and are easily reformatted and displayed in any number of programs. Finally, text files are also in human-readable form, making them ideal for archival purposes.

Unfortunately, text files provide limited flexibility for moving large amounts of data because of the need to convert values into their machine representations, storage and retrieval efficiency, and rounding problems. The additional time required for conversion can be considerable for large datasets.

To read text files into R, use the read.table command. For example, to import a comma-separated values (CSV, usually with suffix ".csv") file that contains row names in the first column, use

```
> fia.plots <- read.table("../../data/fia_plots.csv", sep = ",",
+                         header = TRUE, row.names = 1)
```

The default action of the read.table command is to convert character fields into factors. If you want the characters to remain as characters, then provide as.is = TRUE in the argument list. This inclusion is handy for when you are importing files that contain comments.

A refined version of the read.table function, called read.csv, can also be used to read CSV files. The only difference between the two functions is that the default arguments for read.csv are designed to simplify reading CSV files. Note that here we will use relative directory paths; absolute paths are also supported. Also note that the directories are delimited by forward slashes, even in the Microsoft Windows environment.

```
> fia.plots <- read.csv("../../data/fia_plots.csv")
```

The object that is returned by the **read** family of functions is a data frame object. For example,

```
> class(fia.plots)
```

```
[1] "data.frame"
```

The data frame is one of the fundamental data structures within R and will appear frequently in the functions we will use throughout this book.

The command that we use to save an R table or table-like object to a text file is almost identical to the **read.table** function. To write a data frame object to a text file called fia-plots.csv, we use

```
> write.table(fia.plots, file = "fia-plots.csv")
```

The names of these two functions are slightly misleading since the functions can read and write data structures other than tables.

The **read.fwf** function reads data in fixed-width format. This is useful for, among other tasks, reading data that have been formatted for applications written in FORTRAN. The **read.fwf** function accepts an argument that reports the width of individual data fields. The **widths** argument accepts an integer vector of fixed-width field lengths or a list of integer vectors giving widths for multi-line records.

For example, to read a Forest Vegetation Simulator (FVS, see Wykoff et al., 1982) input tree file

```
1  1     161PP 54.4    165        8
1  2     161DF 16.4    101        8
2  1      01WF 36.3    171        7
2  2     161PP 35.6    131        5

                          ⋮

3 44     161IC  5.5     45        7
3 45     161IC 10.8     53        7
4  1     161DF 14.6     91        8
4  2     161WF  9.6     83        9
4  3     161PP 46.8    159        6
4  4     161M   8.2     61        6
```

using the format defined at the FVS web site, we can use the **read.fwf** function as follows:

```
> fvs.trees <-
+   read.fwf("../../data/stnd73.fvs",
+           widths = c(4, 3, 6, 1, 3, 4, -3, 3, -7, 1),
+           as.is = FALSE, row.names = NULL,
+           col.names = c("plot", "tree", "tree.count",
+             "history", "species", "dbh", "live.tht",
+             "crown.code"))
```

The negative numbers represent portions of the file to be skipped. We might then produce a plot of the heights against diameters by species (Figure 2.1) using Deepayan Sarkar's xyplot function in the lattice package (Sarkar, 2010),

```
> library(lattice)
```

```
> xyplot(live.tht ~ dbh | species, data = fvs.trees)
```

The final data import function that we discuss is scan. The scan function provides the ability to examine data line by line from a file, the console, or a connection. This ability is useful when reading files that have complex structure; for example, when data and metadata are combined. A specific example is when tree-specific and plot-specific information are combined in one file for a number of plots.

The output of scan is a list object in which each row of the file is stored as a separate object. We would then use a loop to process the list. Inside the loop, we would use the substr function to examine specific portions of the row in order to decide what to do with it.

A simple example follows. We have three plots upon which trees were measured for species and diameter by two different crews. The dataset is

```
Plot 001  Crew A
1     DF    23.7
2     GF    40.1
Plot 002  Crew A   Clearfell
Plot 003  Crew B
1     GF    122.6
2     GF    20.3
```

We import the data using the following code:

```
> eg <- scan(file = "../../data/scan-example.txt",
+           sep = "\n", what = "")
```

Note that by using sep = "\n" we ask R to use the end of line to delimit the input chunks, and by using the what = "" argument we instruct R to read each chunk (row) as a character string.

We now create two empty lists to contain the plot and the tree information. We know that the number of trees and plots must be less than the length of the scanned object.

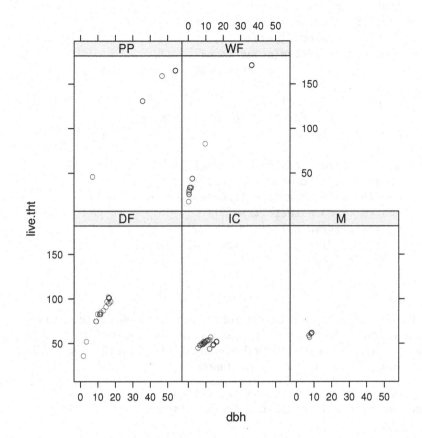

Fig. 2.1: DBH–height plot of data from the FVS file (`stnd73.fvs`) created using the `xyplot` function.

```
> n.in <- length(eg)
> eg.trees <- eg.plots <- vector(mode = "list", length = n.in)
> plot.n <- tree.n <- 1
```

We use a different approach to processing the chunk depending on whether it is a plot or a tree. If it is a plot, we extract portions of the string and save them as variables. If it is a tree, then we split the string on whitespace, which creates a list object, and select the portions of the list that are of interest to us. The split is handled by the `strsplit` function, which accepts regular expressions (regex) as splitting criteria; " +" means "one or more instances of whitespace".

```
> for (i in 1 : n.in) {
+    chunk <- eg[[i]]
```

```
+    if (substr(chunk, 1, 4) == "Plot") {
+        plot.id <- as.numeric(substr(chunk, 6, 9))
+        crew.id <- substr(chunk, 16, 16)
+        comments <- ifelse(nchar(chunk) > 17,
+                          substr(chunk, 17, nchar(chunk)),
+                          "")
+        eg.plots[[plot.n]] <-
+                list(plot.id, crew.id, comments)
+        plot.n <- plot.n + 1
+    } else {
+        tree <- strsplit(chunk, " +")[[1]]
+        tree.id <- as.character(tree[1])
+        species <- as.character(tree[2])
+        dbh.cm <- as.numeric(tree[3])
+        eg.trees[[tree.n]] <-
+                list(plot.id, tree.id, species, dbh.cm)
+        tree.n <- tree.n + 1
+    }
+ }
```

Also, the plot identification for the tree record is *retained* from the most recent iteration of processing the plot information. We conclude by forming, naming, and examining the plot and tree objects. We form the objects from the lists using the do.call and rbind functions.

```
> eg.plots <- as.data.frame(do.call(rbind, eg.plots))
> names(eg.plots) <- c("plot", "crew", "comments")
> eg.plots

  plot crew    comments
1    1    A
2    2    A   Clearfell
3    3    B

> eg.trees <- as.data.frame(do.call(rbind, eg.trees))
> names(eg.trees) <- c("plot", "tree", "species", "dbh.cm")
> eg.trees

  plot tree species dbh.cm
1    1    1      DF   23.7
2    1    2      GF   40.1
3    3    1      GF  122.6
4    3    2      GF   20.3
```

Our example code is quite inefficient. For example, we could handle the conversions of the data types outside the loops. However, it will suffice for the purposes of demonstrating the processing steps.

The `scan` function is very powerful and flexible, and our example has barely scratched the surface of its use. For example, much larger files can be handled by using the `skip` and `nlines` arguments.

Next we examine reading the second most common data format used in forest resource analysis: spreadsheets.

2.2.2 Spreadsheets

In forestry, as in most fields, spreadsheets are extremely common as data storage and communication devices. However, there are shortcomings that limit the spreadsheet's utility for these purposes. Most spreadsheet applications cannot separate the data from the analysis routines and formulas, so that performing a new analysis requires an entirely new spreadsheet, which may involve cutting and pasting cell formulas, etc. Some spreadsheets can contain many subsheets, and the subsheets can all contain inter-connected formulas, macros, and so on. Not all programs can read multiple-sheeted spreadsheets, and some applications may be confused by other contents within the file. Most spreadsheet programs are not platform independent, which makes them difficult to use in a distributed environment or a web-based setting. Finally, some spreadsheets are encoded in proprietary formats that encrypt the data and any associated analysis unnecessarily.

For these reasons, we recommend that spreadsheets not be used for maintaining data. While spreadsheets have their place for simple data-editing tasks and for prototyping analysis methods, we advocate that the storage and management of data be in text files for small projects and in formal databases for large projects. The data should be extracted from the file or the database using sequences of commands that are recorded in documented scripts. We recommend that data be stored in a relational database management system (RDMS) such as PostgreSQL, MySQL, Microsoft Access, or Oracle, and that the RODBC package (Section 2.2.3) be used for data retrieval and updates.

In some circumstances, there may be no way to avoid the use of spreadsheets for data storage. If so, Windows users can use `odbcConnectExcel` in the RODBC package, originally developed by Michael Lapsley and now maintained by Brian Ripley (Ripley and Lapsley, 2009). The `odbcConnectExcel` function can select and extract rows and columns from any of the sheets in an Excel spreadsheet file. This approach allows data importation into R, but it does not impede or interrupt the disparate, or incongruous, data cycle present in many organizations (Brackett, 2000).

2.2.3 Using SQL in R

The Structured Query Language (SQL) is a standard language for writing commands that can be used to access data within RDMS such as Oracle, Microsoft SQL Server, and PostgreSQL (Kline and Kline, 2001). Using an RDMS with SQL for data storage and retrieval provides efficient, straightforward, and consistent interface to data in both simple or complex arrangements.

For example, suppose we have a table in a database named plots. Using one of the database accessibility packages such as RODBC, RPostgreSQL, or RMySQL to access the data stored within the RDMS, the SQL select string to select and retrieve all the data associated with the plots table would be the same. Using the RPostgreSQL package (Prayaga et al., 2009), for example, the resulting code snippet to retrieve data from the plots table would look like

```
> library(RPostgreSQL)
> drv <- dbDriver("PostgreSQL")
> con <- dbConnect(drv,
+                   dbname="forestco",
+                   user="hamannj",
+                   host="localhost")
> sql.command <-
+   sprintf( "select * from plots where plottype = 'fixed';" )
> rs <- dbSendQuery(con, statement = sql.command )
> fixed.plots <- fetch(rs, n = -1)
> dbDisconnect(con)
```

where the resulting object from the `fetch` function, `fixed.plots`, contains a data frame object that contains all plots where the `plottype` is fixed.

In this book, however, we do not present data access methods for relational databases using SQL since tasks commonly associated with databases use large datasets, and project management with large datasets is beyond the scope of this book. For more complete descriptions and examples of accessing an RDMS, see any number of database access libraries, e.g., RODBC, RPostgreSQL, and RMySQL (James and DebRoy, 2009) and the R-data document that comes with the R documentation.

2.2.4 The foreign Package

R provides a package called foreign to read and write files in formats that are used by other software tools (R core members et al., 2010). The foreign package is a collection of file translation functions that can be used to read and write various file formats of common statistical and database applications. Table 2.1 presents the functions for reading and writing data currently available in the foreign package.

Table 2.1: A list of functions available for reading and writing common file formats.

Function	Purpose
data.restore	Read an S3 binary file
lookup.xport	Look up information on a SAS XPORT format library
read.dbf	Read a DBF file
read.dta	Read Stata binary files
read.epiinfo	Read Epi Info data files
read.mtp	Read a Minitab Portable Worksheet
read.octave	Read Octave text data files
read.S	Read an S3 binary File
read.spss	Read an SPSS data file
read.ssd	Obtain a data frame from a SAS permanent dataset via read.xport
read.systat	Obtain a data frame from a Systat file
read.xport	Read a SAS XPORT format library
write.dbf	Write a DBF file
write.dta	Write files in Stata binary format
write.foreign	Write text files and code to read them

Most of the functions in the foreign package are simple to use. For example, the code to read a dBase-formatted file (e.g., **stands.dbf**) using the foreign package is

```
> library(foreign)
> stands <- read.dbf("../../data/stands.dbf")
```

As before, it is useful to immediately examine the imported object. We learn something about the object by using the **names** function (we would also use **str**, but doing so here is a poor use of paper!).

```
> names(stands)
 [1] "SP_ID"       "AREA"        "PERIMETER"  "STANDID"
 [5] "ALLOCATION"  "TAGE"        "BHAGE"      "DF_SITE"
 [9] "TOTHT"       "CUBVOL_AC"   "TPA"        "QMD"
[13] "BA"
```

We can then plot the count of stands in hexagonal bins of quadratic mean diameter and tree counts per hectare using the **plot** and the **hexbin** functions, the latter from the hexbin package (Carr et al., 2010) (Figure 2.2).

```
> stands.non.zero <- stands[stands$QMD > 0,]
> plot(hexbin(stands.non.zero$QMD*2.54 ~
```

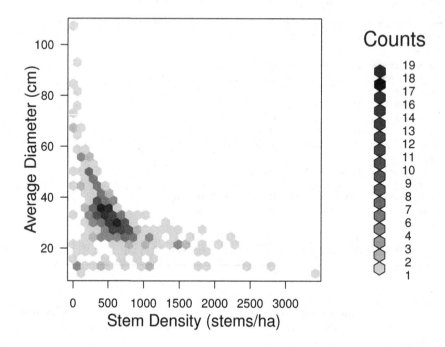

Fig. 2.2: A hexbin plot of stand counts in classes of average diameter and tree density.

```
+                    stands.non.zero$TPA*2.47),
+        ylab = "Average Diameter (cm)",
+        xlab = "Stem Density (stems/ha)")
```

which looks like many other size–density plots presented in the literature (Reineke, 1933; Drew and Flewelling, 1979; Curtis, 1982).

2.2.5 Geographic Data

Several R packages enable reading and writing geographic data in the format of some of the more popular GIS applications, such as Arc/Info and GRASS. We will examine the functions that read ESRI shapefiles that are provided by Nicholas J. Lewin-Koh and Roger Bivand's maptools package (Lewin-Koh and Bivand, 2010). Packages to read and write other GIS file formats can be found at CRAN.

Shapefiles can contain information about many individual polygons. If the shapefiles are relatively small and only consist of a few hundred polygons, then R is suitable for making thematic maps, performing analyses, and storing and updating data within the shapefiles. However, R should not be considered a substitute for a fully featured GIS.

We begin by loading a shapefile of the forest inventory that is made available by Oregon State University[1]. (**stands.shp**). First, we start by loading the package and reading the shapefile using the **readShapePoly** command

```
> library(maptools)
> stands <- readShapePoly("../../data/stands.shp")
```

The total area, in acres, is

```
> sum( stands$AREA ) / 43560.0
```

```
[1] 11994.71
```

and the number of stands is

```
> nrow(stands)
```

```
[1] 510
```

We use the **names** function to obtain the names of the attributes:

```
> names(stands)
```

```
 [1] "SP_ID"      "AREA"        "PERIMETER"  "STANDID"
 [5] "ALLOCATION" "TAGE"        "BHAGE"      "DF_SITE"
 [9] "TOTHT"      "CUBVOL_AC"   "TPA"        "QMD"
[13] "BA"
```

Then, we can use the **plot** command to produce a simple map such as that shown in Figure 2.3

```
> plot(stands, axes = TRUE)
```

2.2.6 Other Data Formats

The set of functions we presented in the previous sections was not exhaustive but covers the bulk of the formats in which forestry data tend to be stored. There are additional file management topics such as how to format a file

---

[1] These data can be found at http://www.cof.orst.edu/cf/gis/.

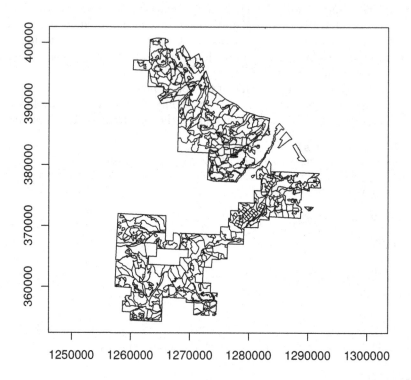

Fig. 2.3: A plot of the ESRI shapefile that contains stand data for the dataset presented in Section 2.4.6.

into single or multiple time series objects, constructing geospatial data from scratch, how to read and write various files for linear programming solvers, and reading remotely sensed images from the Internet. These tasks are beyond the scope of this book. In future chapters, when we make use of an additional file format or reformatting function, we will simply apply the function in the package we happen to be using and suggest the reader become familiar with the documentation.

2.3 Data Management Functions

Most of the data we use for our projects were collected with a specific question, although often one that is quite unlike the question we are examining. In fact, it is not unusual that a completely different analysis project was in mind when the data that we must use were originally collected. Most likely, the data were collected under less than ideal conditions and without clear protocols. Each analysis project usually requires that some effort be spent on reformatting and checking data. This section is designed to introduce readers to a few data manipulation and management functions in R. At the time of writing, the authors have, between them, about 40 years of experience in handling forestry data and have never worked with a dataset that did not demand some level of scrutiny for structure and cleanliness.

2.3.1 Herbicide Trial Data

The herbdata dataset started as a herbicide trial. The plots were installed during the 1994 planting season in southwestern Washington by Don Wallace and Bruce Alber. Three replications of 20 seedlings were planted in two blocks. The two blocks were a control block and a block treated with 220 ml per hectare of Oust herbicide (DuPont, 2005). The plots were then measured over the next ten years. At each observation, the basal diameter, total height, and condition of the stem were recorded. When the stems reached breast height (1.37 m in the United States), the breast height diameter was also recorded. An indicator variable was used to record if the stem was dead or alive. If the stem was dead, the observations were recorded as NA. Recall that to read in the data from a text file, we use the read.table function

```
> herbdata <- read.table("../../data/herbdata.txt",
+                         header = TRUE, sep = ",")
```

and to make sure we have suitable data by printing the first five rows of data using the str function.

```
> str(herbdata)
'data.frame':          960 obs. of  8 variables:
 $ treat  : Factor w/ 2 levels "CONTROL","OUST": 1 1 ...
 $ rep    : Factor w/ 3 levels "A","B","C": 1 1 ...
 $ tree   : int  1 1 ...
 $ date   : Factor w/ 8 levels "10/7/1996 0:00:00",..: 1 2 ...
 $ isalive: int  1 1 ...
 $ height : num  31 59 ...
 $ dia    : num  6.5 9 ...
 $ dbh    : num  0 0 ...
```

The measurement dates, but not the measurement times, were recorded for the observations. So, we can use the strptime function to reformat the dates, thus removing the time portion of the date (otherwise when we plot the data by date, the times will be printed as YYYY-MM-DD 0:00:00) and storing the resulting objects using a special object class that represents temporal data, POSIXct.

```
> herbdata$date <- as.POSIXct(strptime(herbdata$date,
+                                       "%m/%d/%Y"))
```

We can then plot the data (Figure 2.4) using the `coplot` function, which creates conditioning plots:

```
> coplot(height ~ dia | treat * rep, type = "p",
+        data = herbdata[herbdata$isalive == 1,],
+        ylab = "Height (cm)", xlab = "Basal Diameter (mm)")
```

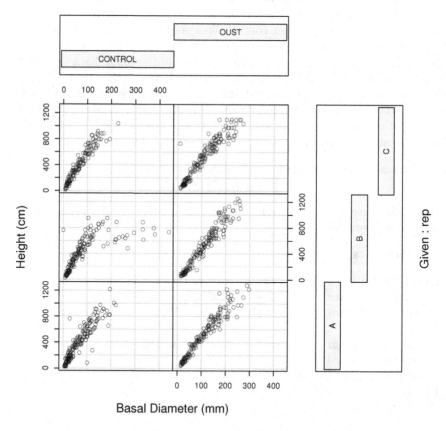

Fig. 2.4: A conditioning plot of the observations in the herbdata data frame.

Something looks suspicious in the figure. The cloud on the right side of the control plot in the B replication looks out of place. We suspect measurement or process error in the data, and we should investigate further, as follows.

2.3.2 Simple Error Checking

Making use of simple tables and graphics to check for outliers and invalid values in datasets is common practice. To produce a summary of the variables in a data frame object, use the summary function:

```
> summary(herbdata)
```

```
    treat       rep              tree
 CONTROL:480   A:320   Min.    : 1.00
 OUST   :480   B:320   1st Qu.:15.75
               C:320   Median :30.50
                       Mean   :30.50
                       3rd Qu.:45.25
                       Max.   :60.00
```

```
         date                    isalive            height
 Min.   :1996-10-07 00:00:00   Min.   :0.0000   Min.   :   15.0
 1st Qu.:1999-03-09 18:45:00   1st Qu.:1.0000   1st Qu.:  193.0
 Median :2001-10-03 23:30:00   Median :1.0000   Median :  449.9
 Mean   :2001-04-30 11:37:30   Mean   :0.9615   Mean   :  472.1
 3rd Qu.:2003-04-04 05:15:00   3rd Qu.:1.0000   3rd Qu.:  693.4
 Max.   :2005-08-23 00:00:00   Max.   :1.0000   Max.   : 1280.2
                                                NAs    :   37.0
```

```
      dia              dbh
 Min.   :  0.00   Min.   :  0.00
 1st Qu.: 36.75   1st Qu.:  0.00
 Median : 82.00   Median : 22.00
 Mean   : 95.12   Mean   : 52.96
 3rd Qu.:138.50   3rd Qu.: 98.00
 Max.   :439.42   Max.   :342.90
 NAs    : 37.00   NAs    : 37.00
```

We see that some of the observations contain NA values, which signify missing values. Other than that, nothing immediately appears to be wrong with the data. We display the first few observations to assess whether or not the other observations in those rows that include NA entries are correct:

```
> head(herbdata[is.na(herbdata$height),])
```

```
      treat rep tree       date isalive height dia dbh
 282 CONTROL   B   36 1997-11-11       0     NA  NA  NA
```

```
283 CONTROL   B   36 1999-08-18        0     NA  NA  NA
284 CONTROL   B   36 2001-02-07        0     NA  NA  NA
285 CONTROL   B   36 2002-05-31        0     NA  NA  NA
286 CONTROL   B   36 2002-11-26        0     NA  NA  NA
287 CONTROL   B   36 2004-04-26        0     NA  NA  NA
```

We suspect that the trees with missing measurements are those trees that are not alive (isalive = 0). This conjecture can be verified by counting the rows that have missing values anywhere for each value of isalive, as follows.

```
> table(complete.cases(herbdata), herbdata$isalive)

           0    1
  FALSE   37    0
  TRUE     0  923
```

As we suspected, the rows that contain missing values, which correspond to the complete.cases function returning FALSE, are all dead trees. To complete our scrutiny, in the following section we will visually inspect the relationships among the variables.

2.3.3 Graphical error checking

Graphical error-checking methods are straightforward using basic plotting functions. As we saw in Figure 2.4, graphical error checking provided quick identification for the suspect observations using a conditioning plot (Chambers, 1991; Cleveland, 1993).

We created a conditioning plot by plotting the height–diameter scatterplots, conditioned on treatment and replication (Figure 2.4). Inspection of Figure 2.4 reveals outliers in the B replication of the control group. In order to focus on these outliers we now need to produce a conditioning plot by measurement date (Figure 2.5):

```
> coplot(height ~ dia | treat * factor(date),
+        data = herbdata[herbdata$isalive == 1,],
+        type = "p",
+        ylab = "Height (cm)",
+        xlab = "Basal Diameter (mm)")
```

We can see that the measurements taken on 2004-04-26 contain the suspicious observations. We can conveniently narrow in on those observations using index operators. To get the correct levels of the factors, use the levels function, which returns the values of the levels of its argument.

```
> levels(herbdata$treat)
```

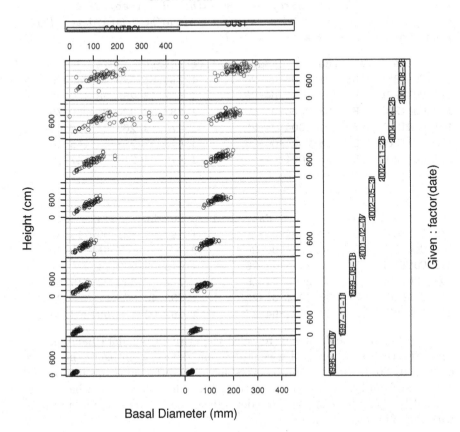

Fig. 2.5: A conditioning plot, by measurement date, of the observations in the herb-data data frame.

```
[1] "CONTROL" "OUST"

> levels(herbdata$rep)

[1] "A" "B" "C"
```

The date variable is not a factor, so we handle it differently

```
> sort(unique(herbdata$date))

[1] "1996-10-07 EST" "1997-11-11 EST" "1999-08-18 EST"
[4] "2001-02-07 EST" "2002-05-31 EST" "2002-11-26 EST"
[7] "2004-04-26 EST" "2005-08-23 EST"
```

so we can use the values to index those problematic observations,

```
> bad.index <- herbdata$treat == levels(herbdata$treat)[1] &
+               herbdata$rep == levels(herbdata$rep)[2] &
+               herbdata$date == sort(unique(herbdata$date))[7]
> bad.data <- herbdata[bad.index,]
```

	treat	rep	tree	date	isalive	height	dia	dbh
167	CONTROL	B	21	2004-04-26	1	734.56	439.42	342.90
175	CONTROL	B	22	2004-04-26	1	637.03	259.08	187.96
183	CONTROL	B	23	2004-04-26	1	883.92	330.20	266.70
191	CONTROL	B	24	2004-04-26	1	478.53	269.24	157.48
199	CONTROL	B	25	2004-04-26	1	646.17	233.68	165.10
207	CONTROL	B	26	2004-04-26	1	612.64	218.44	147.32

$$\vdots$$

	treat	rep	tree	date	isalive	height	dia	dbh
279	CONTROL	B	35	2004-04-26	1	697.99	381.00	213.36
287	CONTROL	B	36	2004-04-26	0	NA	NA	NA
295	CONTROL	B	37	2004-04-26	1	615.69	332.74	231.14
303	CONTROL	B	38	2004-04-26	1	774.19	330.20	279.40
311	CONTROL	B	39	2004-04-26	1	792.48	375.92	330.20
319	CONTROL	B	40	2004-04-26	1	762.00	0.00	203.20

A subsequent telephone call revealed the error in our data handling. In this case, the field crew had used a different device to measure the stems on 2004-04-26. During our unit conversion process, the value was incorrectly converted from imperial units to metric units. The values should have been divided by 2.54. We can use the index values again to correct the problem with the following commands:

```
> herbdata$dia[bad.index] <- herbdata$dia[bad.index] / 2.54
```

Note that we have to correct the breast-height diameter as well:

```
> herbdata$dbh[bad.index] <- herbdata$dbh[bad.index] / 2.54
```

These data-cleaning commands represent important intelligence about the data handling. It is very important that the newly cleaned data object not be saved back over the original dataset. Instead, either the data should be kept as they are and the code that was used to clean the data should be run routinely when the data are used, or a new version of the data, time-stamped, should be saved, and the differences between the original and the cleaned data should be carefully documented in a data dictionary or journal, which should be readily available with the dataset.

Now that the data have been corrected, we may want to revisit our original plots using coplot or Deepayan Sarkar's lattice functions to generate the conditioning plot again, verify our results, and continue with our analysis.

2.3.4 Data Structure Functions

Although we may be able to simply pass an unaltered data frame object to other functions for sample processing, model development, or optimization, we may want to change the structure of the data frame object to facilitate specific programming needs. The main functions for data manipulation are the **split**, **aggregate**, and **merge** functions.

The first of these functions that we examine is the **split** function, which divides a data frame (or any vector) into a list of the groups, distinguished by corresponding with common values of some categorical variable. A *list* is a specific class of R object that is a collection of objects that can be unalike. The reverse operation can be performed by the **unsplit** function.

Imagine that we want to find out how many trees we have in each of the treatments of the herbdata object. Using the **split** function, we can split the data frame into a list of data frames where the name of each treatment is now the item in the list,

```
> split.herb <- split(herbdata, herbdata$treat)
> class(split.herb)

[1] "list"

> names(split.herb)

[1] "CONTROL" "OUST"

> nrow(split.herb$CONTROL)

[1] 480

> nrow(split.herb$OUST)

[1] 480
```

We can also check the names of the members,

```
> names(split.herb$CONTROL)

[1] "treat"    "rep"       "tree"     "date"      "isalive" "height"
[7] "dia"      "dbh"
```

A more compact way to obtain the row count information is to use the powerful **lapply** function, which applies its second argument as a function to each of the items in the list that is identified as its first argument; for example

```
> lapply(split.herb, nrow)
```

```
$CONTROL
[1] 480

$OUST
[1] 480
```

The `lapply(X, FUN, ...)` function returns a list of the same length as X. Using the newly split data frame, which is stored as a list, we can write `for` loops to process our data or continue to manipulate our data frame for processing using the core data management functions.

For simplicity's sake, we examine a reduced data frame, including only columns 1 (treatment), 2 (replication), 6 (height), and 7 (basal diameter), and print some basic summaries using the `lapply` function:

```
> herbdata.short <- herbdata[,c(1,2,6,7)]
> split.herb.short <- split(herbdata.short, herbdata$treat)
> lapply(split.herb.short, summary)

$CONTROL
     treat        rep         height              dia
 CONTROL:480   A:160   Min.    :   15.0   Min.    :   0.00
 OUST   :  0   B:160   1st Qu.:  175.6   1st Qu.:  26.75
               C:160   Median :  391.4   Median :  62.00
                       Mean    :  413.4   Mean    :  69.55
                       3rd Qu.:  615.7   3rd Qu.: 102.00
                       Max.    : 1219.2   Max.    : 228.60
                       NAs     :   37.0   NAs     :  37.00

$OUST
     treat        rep         height              dia
 CONTROL:  0   A:160   Min.    :   42.0   Min.    :  10.0
 OUST   :480   B:160   1st Qu.:  212.6   1st Qu.:  51.0
               C:160   Median :  530.4   Median : 109.0
                       Mean    :  526.2   Mean    : 112.6
                       3rd Qu.:  762.0   3rd Qu.: 159.0
                       Max.    : 1280.2   Max.    : 298.4
```

The `aggregate` function is useful for collapsing data into forms familiar to foresters such as stand tables, log-stock tables, and species-level summaries. For example, we might want to know the diameter distribution of the trees by treatment, ignoring the replications, for the herbicide trial data for the latest sample date. We can combine all three functions in conjunction with the ability to select a subset of the measurements by date. We check what dates we have to choose from by printing the unique values, sorted:

```
> sort(unique(herbdata$date))
```

```
[1] "1996-10-07 EST" "1997-11-11 EST" "1999-08-18 EST"
[4] "2001-02-07 EST" "2002-05-31 EST" "2002-11-26 EST"
[7] "2004-04-26 EST" "2005-08-23 EST"
```

To generate our summaries, we select and split only the last measurement:

```
> herbdata.shorter <-
+       herbdata[herbdata$date == max(herbdata$date), c(1,2,6,8)]
> split.herb.shorter <-
+       split(herbdata.shorter, herbdata.shorter$treat)
```

We can now classify the trees by diameter class. We bind that column to a temporary data frame to facilitate the process. We can then use the **aggregate** function to generate totals for each of the treatments by diameter class:

```
> rt <- cbind(herbdata.shorter,
+             dc = cut(herbdata.shorter$dbh,
+             breaks = c(0, 50, 100, 150, 200, 300, 400, 999),
+             labels = c("000--050", "050--100", "100--150",
+                 "150--200", "200--300", "300--400","400+")))
> st <- aggregate(x = list(basal.area = pi/(4*10^2) * rt$dbh^2,
+                          tht = rt$height,
+                          stems = rep(1, nrow(rt))),
+                 by = list(treat = rt$treat,
+                           diac = rt$dc),
+                 FUN = sum)
> st
```

```
    treat     diac basal.area      tht stems
1 CONTROL 000--050   27.61556  2255.52     5
2 CONTROL 050--100  993.34934 13441.68    18
3    OUST 050--100   73.16856   883.92     1
4 CONTROL 100--150 3028.48926 21244.56    26
5    OUST 100--150 2439.45448 17465.04    18
6 CONTROL 150--200 1404.13710  6035.04     6
7    OUST 150--200 9093.28375 40660.32    39
8    OUST 200--300  681.82558  2529.84     2
```

From our results, we can see that the total height value is a sum of all the heights. We really wanted the mean height for the treatment, so we can use the assignment operator (<-) to reassign the **tht** variable to the correct value, now expressed in meters, and print the results again:

```
> st$tht <- st$tht / st$stems / 100
> st
```

Table 2.2: OUST herbicide trials.

Treatment	Dia. Class (mm)	Basal Area (mm²)	Mean Total Height (m)	Stems
CONTROL	000–050	27.616	4.5110	5
CONTROL	050–100	993.349	7.4676	18
CONTROL	100–150	3028.489	8.1710	26
CONTROL	150–200	1404.137	10.0584	6
OUST	050–100	73.169	8.8392	1
OUST	100–150	2439.454	9.7028	18
OUST	150–200	9093.284	10.4257	39
OUST	200–300	681.826	12.6492	2

```
    treat       diac basal.area         tht stems
1 CONTROL 000--050   27.61556  4.511040     5
2 CONTROL 050--100  993.34934  7.467600    18
3    OUST 050--100   73.16856  8.839200     1
4 CONTROL 100--150 3028.48926  8.170985    26
5    OUST 100--150 2439.45448  9.702800    18
6 CONTROL 150--200 1404.13710 10.058400     6
7    OUST 150--200 9093.28375 10.425723    39
8    OUST 200--300  681.82558 12.649200     2
```

Now that we have a summary table, we could change the column names and print a nice-looking table in our LATEX document. Since we are using the Sweave/Rtangle functions to generate this book, we can create a table using the latex function after sorting the results (Table 2.2).

```
> cap <- "OUST herbicide trials."
> st <- st[order(st$treat, st$diac),]
> st$treat <- as.character(st$treat)
> st$diac <- as.character(st$diac)
> names(st) <- c("Treatment", "Dia. Class (mm)",
+                "Basal Area ($\\mbox{mm}^2$)",
+                "Mean Total Height (m)", "Stems")

> latex(st, rowlabel = NULL, rowname = NULL, file="",
+       caption = cap, label = "tab:herbdata_results",
+       digits = 5, booktabs = TRUE,
+       cjust = c("l","c","r","r","r"))
```

Finally, the merge function merges two data frames by common columns or row names. This function can perform one-to-one, many-to-one and many-to-many merges. By default, the data frames are merged on common column names, but separate specifications of the columns can be given by the by.x and by.y arguments, using either name or index number. The merge function can be used to append columns onto a data frame where y is the result of summarization operations at various levels such as plots or clusters.

For example, imagine that we wish to fit a height–diameter model for the herbicide trials and we want to include the plot-level total basal areas in our list of potential regressors as a measure of local competition. For this exercise, we will not concern ourselves with model specification. We are interested simply in demonstrating the usefulness of the merge function.

We know that we have two blocks in the herbicide trial data

```
> names(split.herb)
```

```
[1] "CONTROL" "OUST"
```

and we want to construct a data frame that contains the total basal diameter area and breast height (1.3 m) area on each replication for each time period, treatment and replication. Note that we use the with function, which attaches the object names in the first argument to the search path while it executes the code identified in the second argument. It is no more efficient but produces code that is easier to read.

```
> areas <-
+    with(herbdata,
+         aggregate(x = list(plot.bh.area = pi/400 * dbh^2,
+                            plot.bas.area = pi/400 * dia^2),
+              by = list(treat = treat,
+                        rep = rep,
+                        date = date),
+              FUN = sum))
```

To verify our results, we will print the first ten rows of the resulting data frame:

```
> areas[1:10,]
```

	treat	rep	date	plot.bh.area	plot.bas.area
1	CONTROL	A	1996-10-07	0	61.01271
2	OUST	A	1996-10-07	0	86.17891
3	CONTROL	B	1996-10-07	0	49.58501
4	OUST	B	1996-10-07	0	98.32062
5	CONTROL	C	1996-10-07	NA	NA
6	OUST	C	1996-10-07	0	98.47361
7	CONTROL	A	1997-11-11	0	128.76603
8	OUST	A	1997-11-11	0	236.90554
9	CONTROL	B	1997-11-11	NA	NA
10	OUST	B	1997-11-11	0	293.01438

Something does not look correct because there are basal diameter values that contain NAs and we know that not all of the stems died within those blocks on those measure dates. Since we have used the default arguments for the sum function, the missing diameter observations have caused the sum function to return NA. We have to pass the `na.rm = TRUE` argument to sum to tell it to ignore the missing values:

```
> areas <-
+    with(herbdata,
+        aggregate(x = list(plot.bh.area = pi/400 * dbh^2,
+                           plot.bas.area = pi/400 * dia^2),
+                  by = list(treat = treat,
+                            rep = rep,
+                            date = date),
+                  FUN = sum,
+                  na.rm = TRUE))
> areas[1:10,]
      treat rep       date plot.bh.area plot.bas.area
1   CONTROL   A 1996-10-07            0      61.01271
2      OUST   A 1996-10-07            0      86.17891
3   CONTROL   B 1996-10-07            0      49.58501
4      OUST   B 1996-10-07            0      98.32062
5   CONTROL   C 1996-10-07            0      41.80674
6      OUST   C 1996-10-07            0      98.47361
7   CONTROL   A 1997-11-11            0     128.76603
8      OUST   A 1997-11-11            0     236.90554
9   CONTROL   B 1997-11-11            0     105.53984
10     OUST   B 1997-11-11            0     293.01438
```

We can now merge the results based on the values of the original data frame with the new aggregated data frame and print the names of the results to verify the operation worked as expected:

```
> final.data <- merge(herbdata, areas)
> names(final.data)

 [1] "treat"          "rep"           "date"
 [4] "tree"           "isalive"       "height"
 [7] "dia"            "dbh"           "plot.bh.area"
[10] "plot.bas.area"

> head(final.data[,c(1,2,3,4,7,10)])
      treat rep       date tree  dia plot.bas.area
1  CONTROL   A 1996-10-07    1  6.5      61.01271
2  CONTROL   A 1996-10-07    6 24.4      61.01271
3  CONTROL   A 1996-10-07   18 13.4      61.01271
4  CONTROL   A 1996-10-07    3 22.6      61.01271
5  CONTROL   A 1996-10-07    8 26.3      61.01271
6  CONTROL   A 1996-10-07   20 16.6      61.01271
```

We can use this dataset to fit our model. We cover the use of R for hierarchical regression in Chapter 7.

2.4 Examples

We now demonstrate the tools and principles that were developed in the preceding sections to read and scrutinize a number of forestry-related databases that will be used in subsequent chapters of the book.

One operation that we will continuously rely upon is the identification of the counts of missing values in the variables of the datasets. We have written an R function to compactly report this information as follows.

```
> show.cols.with.na <- function(x) {
+    ## First, check that object is a data frame
+    if (class(x) != "data.frame")
+      stop("x must be a data frame.\n")
+    ## Count the missing values by column.
+    missing.by.column <- colSums(is.na(x))
+    ## Are any missing?
+    if (sum(missing.by.column) == 0) {
+      cat("No missing values.\n")
+    } else {
+      ## Only return columns with missing values.
+      missing <- which(missing.by.column > 0)
+      return(missing.by.column[missing])
+    }
+ }
```

This function will report the number of missing observations by variable for only those variables that have missing values.

2.4.1 Upper Flat Creek in the UIEF

In the summer of 1991, a stand examination was made of the Upper Flat Creek unit of the University of Idaho Experimental Forest (UIEF). A square grid of 144 plots was laid out, with north–south inter-plot distance of 134.11 m and east–west inter-plot distance of 167.64 m. A 7.0 m^2/ha BAF variable-radius plot was installed at each plot location. Every tree in the plot was measured for species and diameter at 1.37 m (dbh), recorded in mm. A sub-sample of trees was measured for height, recorded in dm, although it is no longer known how this subsample was selected. The area of the stand was 323.8 ha.

2.4.1.1 Tree Data

These data will require a number of processing steps before they can be used for inventory or modeling analysis. For example, some of the units are inappropriate, some trees need to be excluded, and some empty plots need to be recognized. The data are stored as a single comma-delimited flat file.

```
> ufc.tree <- read.csv("../../data/ufc.csv")
```

We use the str function to examine the structure of the object.

```
> str(ufc.tree)
```

```
'data.frame':          637 obs. of  5 variables:
 $ Plot   : int  1 2 ...
 $ Tree   : int  1 1 ...
 $ Species: Factor w/ 13 levels "","DF","ES","F",..: 1 2 ...
 $ Dbh    : int  NA 390 ...
 $ Height : int  NA 205 ...
```

We now count the missing values by variable.

```
> show.cols.with.na(ufc.tree)
```

```
  Dbh Height
   10    248
```

We note that ten records are missing diameter measures. These ten rows represent empty plots and therefore should not be eliminated thoughtlessly. We will have to deal with them as necessary later in the analysis. Some straightforward manipulations are required to deliver the measures in useful units.

```
> names(ufc.tree) <-
+   c("point","tree","species","dbh.mm","ht.dm")
> ufc.tree$dbh.cm <- ufc.tree$dbh.mm / 10
> ufc.tree$ba.m2 <- ufc.tree$dbh.cm^2 / 40000 * pi
> ufc.tree$height.m <- ufc.tree$ht.dm / 10
```

For the moment, we impute the missing heights using species-specific diameter–height models from Wykoff et al. (1982). This is cold-deck imputation (see Section 4.2.2.2). We have written a vectorized function that applies these models in the original imperial measures, and we have also written accompanying metric wrapper functions. These functions are available as part of the package that accompanies this book.

```
> height.hat <- ht.fvs.ni.m(ufc.tree$species, ufc.tree$dbh.cm)
> missing.hts <- is.na(ufc.tree$height.m)
> ufc.tree$height.m[missing.hts] <- height.hat[missing.hts]
```

We estimate tree-level merchantable volumes using functions provided by
Wykoff et al. (1982). Again, we have written a vectorized function that applies
these models in the original imperial measures and accompanying metric
wrapper functions.

```
> ufc.tree$vol.m3 <-
+       with(ufc.tree, vol.fvs.ni.m3(species, dbh.cm, height.m))
```

2.4.1.2 Plot-Level Data

We now create an object that summarizes the plot-level information. In order
to process the tree-level data, we need to invoke some elements of the design.
The *tree factor* is the design basal area factor (BAF) divided by the tree-level
basal area, and the tree-level volume per unit area is the product of the tree-
level volume and the tree factor, as above. We can proceed with the aggregate
command as there are no empty subsets, although there are some that have
missing values. We allow for these in the sum command by appending the
flag: na.rm = TRUE. That flag is then passed by **aggregate** to the function
that it calls.

```
> ufc.baf.met <- 7
> ufc.tree$tf.ha <- ufc.baf.met / ufc.tree$ba.m2
> ufc.tree$vol.m3.ha <- ufc.tree$vol.m3 * ufc.tree$tf.ha
> ufc.SyRS.data <-
+    aggregate(x = list(vol.m3.ha = ufc.tree$vol.m3.ha),
+              by = list(point = ufc.tree$point),
+              FUN = sum,
+              na.rm = TRUE)
> ufc.SyRS.data$weight <- 1
> str(ufc.SyRS.data)

'data.frame':        144 obs. of  3 variables:
 $ point    : int  1 2 3 4 5 6 7 8 9 10 ...
 $ vol.m3.ha: num  0 63.4 204.6 296.8 309.6 ...
 $ weight   : num  1 1 1 1 1 1 1 1 1 1 ...
```

Note that R automatically recycled the 1 that we assigned to the weight
variable in the ufc.SyRS.data data object until it had the same length as
the other variables contained in that object.

We now have a point-level dataset with the necessary design information
and a variable of interest. We named the object ufc.SyRS.data to reflect the
fact that the inventory was performed using a systematic random sample,
about which more can be found in Section 3.3.2. The spatial structure of the
design is reported in the next section.

2.4.1.3 Spatial Data

We also have spatial information about the sample, obtained from sketch maps that were made by the inventory crews. When we add the spatial information to the data frame, we can use R to get a snapshot of the stand with the additional variables. First, we form the data frame using the locations.

```
> locations <- data.frame(point = 1:144,
+                         north.n = rep(c(12:1),12),
+                         east.n = rep(c(1:12), rep(12,12)))
> locations$north <- (locations$north.n - 0.5) * 134.11
> locations$east <- (locations$east.n - 0.5) * 167.64
```

We combine these datasets using merge, which automatically performs matching using those variables that appear in both datasets.

```
> ufc.SyRS.data <- merge(ufc.SyRS.data, locations)
```

Figure 2.6 is then obtained by the following code:

```
> opar <- par(las=1, pty="s")
> plot(ufc.SyRS.data$east, ufc.SyRS.data$north,
+       type = "n", axes = F,
+       xlim = c(0,max(ufc.SyRS.data$east)+167.64/2),
+       ylim = c(0,max(ufc.SyRS.data$north)+134.11/2),
+       xlab = "West-East (m)", ylab = "South-North (m)",
+       main = expression(paste("Units of 50", m^3, "/ha")))
> axis(1); axis(2)
> grayrange <- range(ufc.SyRS.data$vol.m3.ha)
> text(formatC(ufc.SyRS.data$vol.m3.ha/50,
+               format = "f", digits = 0),
+       x = ufc.SyRS.data$east,
+       y = ufc.SyRS.data$north, cex=1.5,
+       col = gray(1 - (ufc.SyRS.data$vol.m3.ha+200)/800))
> par(opar)
```

Now we can use the processed data for further analysis in later chapters. We have two specific pieces: the tree-level measurements, contained in an object called ufc.tree, and the plot-level information, contained in an object called ufc.SyRS.data.

2.4.2 Sweetgum Stem Profiles

This dataset is a collection of stem measurements of a sample of sweetgum (*Liquidambar styraciflua* L.) trees from Texas in the USA. The data were

Units of 50m³/ha

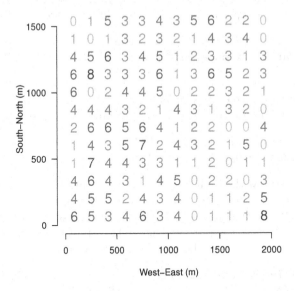

Fig. 2.6: Spatial snapshot of the Upper Flat Creek stand in units of 50 m³/ha.

kindly supplied by Professor Timothy Gregoire and were originally collected by David Lenhart. We will read the data, which are stem profile measures (that is, bole diameters at particular heights), and then use them to compute tree volumes.

The sweetgum measurement data are a good example of the kind of data that require the use of **scan**. The first 26 lines of the file are metadata. The profile measures for each of the 39 trees then follow, each separated by a line that provides information about the tree. Also, the trees each have varying numbers of measurements.

```
> raw.data <- scan("../../data/TX_SGUM2.DAT",
+                   what = "", sep = "\n")
> length(raw.data)

[1] 1101
```

We have 1101 rows of data to work with. Inspection of the raw.data object (not shown here) shows us that the first 26 rows and the last row should be ignored.

```
> raw.data <- raw.data[-c(1:26, 1101)]
```

We need to compare the metadata for the trees to be sure that we understand them and their structure. After examining the file, we see that each tree is consistently identified with the label SWEETGUM. We identify and examine these rows using the grep function in the following code (results not shown).

```
> metadata <- grep("SWEETGUM", raw.data)
> cbind(metadata, raw.data[metadata])
```

This exercise shows us that the plot ID is incomplete for two of the trees. We fix them with educated guesses as follows. The substr function is very powerful: it can be used to report and to alter portions of a string.

```
> substr(raw.data[627], 1, 1) <- "4"
> substr(raw.data[910], 1, 1) <- "5"
```

As is usually true with R, there are numerous ways to process the data. Here we will use brute force to associate plot and tree identifiers with the stem measures. We do this in a loop: every line that is not tree metadata has the first ten characters of the previous line prefixed to it using paste. We can then use these values to merge the tree data and the stem data.

```
> for (i in 1:length(raw.data)) {
+    if(substr(raw.data[i], 57, 64) != "SWEETGUM")
+       raw.data[i] <- paste(substr(raw.data[i - 1], 1, 10),
+                            raw.data[i], sep="")
+ }
```

This quick and dirty solution guarantees that the stem measures are now all associated with the tree data that precede them in the file. We can now extract the tree and section data using the search that we had already performed.

```
> tree.data <- raw.data[metadata]
> length(tree.data)
```

```
[1] 39
```

```
> sections.data <- raw.data[-metadata]
> length(sections.data)
```

```
[1] 1035
```

The variables of interest to us can be extracted directly from these two objects.

```
> sweetgum <-
+     data.frame(plot = factor(substr(tree.data, 1, 5)),
+                tree = substr(tree.data, 6, 10),
+                dbh.in = substr(tree.data, 21, 26),
+                stump.ht.ft = substr(tree.data, 27, 32),
```

```
+                    height.ft = substr(tree.data, 39, 44))
> sections <-
+       data.frame(plot = factor(substr(sections.data, 1, 5)),
+                   tree = substr(sections.data, 6, 10),
+                   meas.ln.ft = substr(sections.data, 11, 16),
+                   meas.dob.in = substr(sections.data, 20, 25),
+                   meas.dib.in = substr(sections.data, 26, 31))
```

We should check the data classes before continuing. We use `sapply` to easily apply the `class` function to each column in the data frames.

```
> sapply(sweetgum, class)
        plot        tree     dbh.in stump.ht.ft   height.ft
    "factor"    "factor"   "factor"    "factor"    "factor"

> sapply(sections, class)
        plot        tree meas.ln.ft meas.dob.in meas.dib.in
    "factor"    "factor"   "factor"    "factor"    "factor"
```

The data are not yet of the appropriate class. We have to convert them. A single loop will suffice.

```
> for (i in 3:5) {
+    sweetgum[,i] <- as.numeric(as.character(sweetgum[,i]))
+    sections[,i] <- as.numeric(as.character(sections[,i]))
+ }
```

We next merge the two data frame objects. R will automatically use the variables that are common to the data frames.

```
> all.meas <- merge(sweetgum, sections, all = TRUE)
> dim(all.meas)

[1] 1035    8

> names(all.meas)

[1] "plot"       "tree"       "dbh.in"      "stump.ht.ft"
[5] "height.ft"  "meas.ln.ft" "meas.dob.in" "meas.dib.in"
```

We now need to convert the data to metric measures for the section data and the tree-level data.

```
> all.meas$meas.ht.ft <- with(all.meas,
+                             meas.ln.ft + stump.ht.ft)
> all.meas$meas.ht.m <- all.meas$meas.ht.ft / 3.2808399
> all.meas$meas.dob.cm <- all.meas$meas.dob.in * 2.54
> sweetgum$height.m <- sweetgum$height.ft / 3.2808399
> sweetgum$dbh.cm <- sweetgum$dbh.in * 2.54
```

Finally, we are in a position to compute the volumes of the trees. We treat
the trees as geometric solids and integrate their sectional areas along their
length. We use a function for this task. This function accepts the diameters
and the heights at which they are measured, along with tree height and lower
limit for volume. It constructs a spline model of the radius as a function of
measurement height. Finally, it computes the sectional area from the output of the spline function evaluated at arbitrary heights and integrates that
quantity along the length of the stem. The argument that supplies the radius
is constrained to be non-negative using the parallel maximum function pmax

```
> spline.vol.m3 <- function(hts.m,
+                           ds.cm,
+                           max.ht.m,
+                           min.ht.m = 0) {
+    rs.cm <- c(ds.cm[order(hts.m)] / 2, 0)
+    hts.m <- c(hts.m[order(hts.m)], max.ht.m)
+    taper <- splinefun(hts.m, rs.cm)
+    volume <- integrate(f = function(x)
+                          pi * (taper(pmax(x,0))/100)^2,
+                        lower = min.ht.m,
+                        upper = max.ht.m)$value
+    return(volume)
+ }
```

We apply this function to the section data and the tree data using the powerful mapply function, along with split.

```
> sweetgum$vol.m3 <-
+    mapply(spline.vol.m3,
+           hts.m = split(all.meas$meas.ht.m, all.meas$tree),
+           ds.cm = split(all.meas$meas.dob.cm, all.meas$tree),
+           max.ht.m = as.list(sweetgum$height.m),
+           min.ht.m = 0.3)
```

We conclude by checking that the volumes are commensurate with our
expectations by comparing the predicted volumes with those of a second-
degree paraboloid with the same height and breast-height cross-sectional area
(Figure 2.7).

```
> par(las = 1)
> plot(sweetgum$vol.m3,
+    (sweetgum$dbh.cm/200)^2 * pi * sweetgum$height.m / 2,
+    ylab = expression(paste("Second-degree paraboloid volume (",
+        m^3, ")", sep="")),
+    xlab = expression(paste("Integrated spline volume (",
+        m^3, ")", sep="")))
> abline(0, 1, col="darkgrey")
```

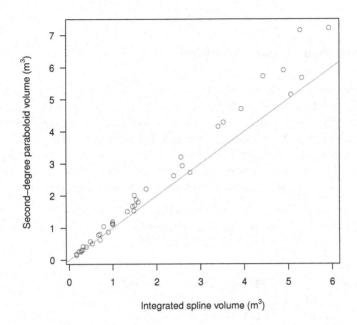

Fig. 2.7: Comparison of sweetgum volumes computed by integrating the sectional area as estimated from a spline fit to the height–radius profile (x-axis) and assuming a second-degree paraboloid with the same height and breast-height cross-sectional area (y-axis).

2.4.3 FIA Data

These data are plot-level measures of stand basal area and unweighted mean tree height from the USDA Forest Service Forest Inventory and Analysis program. The data are an approximately systematic sample of 2632 plots from the Inland Empire of the western United States. More information on the origin and processing of these data can be found in Froese (2003). The geographic location of the plots is obscured; we only know the identifier of the national forest to which the plots are closest. Reading and processing these data is straightforward, largely because we benefit from Froese's (2003) hard work.

```
> fia.plots <- read.csv("../../data/fia_plots.csv")
> fia.plots$forest <- factor(fia.plots$forest)
```

```
> fia.plots$ba.m2.ha <- fia.plots$ba * 2.47105381 / 10.7639104
> fia.plots$ht.m <- fia.plots$ht * 0.3048
```

2.4.4 Norway Spruce Profiles

The Norway spruce (*Picea abies* [L.] Karst) data are tree data originally reported in von Guttenberg (1915) and kindly provided to us by Professor Boris Zeide. The dataset comprises measures taken on 107 average-size trees from seven locations that encompassed five different sites. These data are more fully documented in Zeide (1993).

We read the comma-delimited file, convert the variable names to be all lowercase, change one of them, and examine the resulting object.

```
> gutten <- read.csv("../../data/gutten.csv")
> names(gutten) <- tolower(names(gutten))
> names(gutten)[names(gutten)=="diameter"] <- "dbh.cm"
> str(gutten)
```

```
'data.frame':           1287 obs. of  8 variables:
 $ site    : int  1 1 1 1 1 1 1 1 1 ...
 $ location: int  1 1 1 1 1 1 1 1 1 ...
 $ tree    : int  1 1 1 1 1 1 1 1 1 ...
 $ age.base: int  10 20 30 40 50 60 70 80 ...
 $ height  : num  1.2 4.2 9.3 14.9 19.7 23 25.8 27.4 ...
 $ dbh.cm  : num  NA 4.6 10.2 14.9 18.3 20.7 22.6 24.1 ...
 $ volume  : num  0.3 5 38 123 263 400 555 688 ...
 $ age.bh  : num  NA 9.67 19.67 29.67 ...
```

It would be useful to ensure that we have a unique tree identifier.

```
> gutten$site <- factor(gutten$site)
> gutten$location <- factor(gutten$location)
> gutten$tree.ID <- with(gutten, interaction(location, tree))
```

We can now count the trees within site and location classes by reducing the first three columns of the data frame to their unique components and tabulating them. We use the **unique** function. Here we count trees by location and then by location and site.

```
> with(unique(gutten[,c("site","location","tree.ID")]),
+      table(location))
```

```
location
 1  2  3  4  5  6  7
15  2 13 10 46 11 10
```

```
> with(unique(gutten[,c("site","location","tree.ID")]),
+       table(location, site))
```

```
          site
location  1  2  3  4  5
       1  6  4  5  0  0
       2  2  0  0  0  0
       3  7  4  2  0  0
       4  3  2  2  0  3
       5  3 19  7 15  2
       6  0  7  2  2  0
       7  0  1  2  4  3
```

There seems to be substantial imbalance in the tree counts in the different locations — ranging from two at location 2 to 46 at location 5. Next, we check whether any observations contain missing values.

```
> show.cols.with.na(gutten)
```

```
dbh.cm volume age.bh
    87      6     87
```

The 87 measures with missing diameter and age at breast height were each presumably taken while the tree height was less than or equal to 1.3 m. We can check this conjecture by asking R to report the maximum height associated with all the tree measures that have missing diameters. We identify the values that are missing by using the is.na function.

```
> max(gutten$height[is.na(gutten$dbh.cm)])
```

```
[1] 1.3
```

Our conjecture seems reasonable. We will exclude these measures from the data for the purposes of modeling.

```
> gutten <- gutten[!is.na(gutten$dbh.cm),]
```

These data are subsequently used in Sections 6.2 and 7.6.

2.4.5 Grand Fir Profiles

The grand fir (*Abies grandis* (Dougl.) Lindl.) data were provided to us by Dr. Albert Stage. To give a brief synopsis of the design, a sample of 66 trees was selected in national forests around northern and central Idaho. According to Stage (pers. comm., 2003), the trees were selected purposively rather than randomly. Stage (1963) noted that the selected trees "...appeared to have been dominant throughout their lives" and "...showed no visible evidence

of crown damage, forks, broken tops, etc." The habitat type and diameter
outside bark at 1.37 m (4′6″) were also recorded for each tree, as was the
national forest from which it came. Each tree was then split, and decadal
measures were made of height and diameter inside bark at 1.37 m (4′6″). We
have data from nine national forests and six different habitat types.

We import the data as follows:

```
> stage <- read.csv("../../data/stage.csv")
> str(stage)
```

```
'data.frame':        542 obs. of  7 variables:
 $ Tree.ID: int  1 1 1 1 1 2 2 2 2 2 ...
 $ Forest : int  4 4 4 4 4 4 4 4 4 4 ...
 $ HabType: int  5 5 5 5 5 5 5 5 5 5 ...
 $ Decade : int  0 1 2 3 4 0 1 2 3 4 ...
 $ Dbhib  : num  14.6 12.4 8.8 7 4 20 18.8 17 15.9 14 ...
 $ Height : num  71.4 61.4 40.1 28.6 19.6 ...
 $ Age    : int  55 45 35 25 15 107 97 87 77 67 ...
```

Some cleaning and manipulation will be necessary. We start by defining
the factors.

```
> stage$Tree.ID <- factor(stage$Tree.ID)
> stage$Forest.ID <- factor(stage$Forest, labels = c("Kaniksu",
+     "Coeur d'Alene", "St. Joe", "Clearwater", "Nez Perce",
+     "Clark Fork","Umatilla", "Wallowa", "Payette"))
```

The following habitat codes refer to the climax tree species, which is the
most shade-tolerant species that can grow on the site, and the dominant un-
derstory plant, respectively. Ts refers to *Thuja plicata* and *Tsuga heterophylla*,
Th refers to just *Thuja plicata*, AG is *Abies grandis*, PA is *Picea engelmanii*
and *Abies lasiocarpa*, Pach is *Pachistima myrsinites*, and Op is the nasty
Oplopanaz horridurn.

Grand fir is considered a major climax species for AG/Pach, a major seral
species for Th/Pach and PA/Pach, and a minor seral species for Ts/Pach
and Ts/Op. Loosely speaking, a community is *seral* if there is evidence that
at least some of the species are temporary and *climax* if the community is
self-regenerating (Daubenmire, 1952).

```
> stage$Hab.ID <- factor(stage$HabType, labels = c("Ts/Pac",
+     "Ts/Op", "Th/Pach", "AG/Pach", "PA/Pach"))
```

The measurements are all imperial (this was about 1960, after all). We compute metric measures.

```
> stage$dbhib.cm <- stage$Dbhib * 2.54
> stage$height.m <- stage$Height / 3.2808399
```

A final check for missing values shows us that there are

```
> show.cols.with.na(stage)
```

No missing values.

2.4.6 McDonald–Dunn Research Forest

The McDonald–Dunn Research Forest is located near Oregon State University in Corvallis, Oregon. The forest is approximately 4856 hectares (11995 acres) and is actively managed for student instruction, revenue, research, and recreation.

The forest has been divided into areas of homogeneity (stands), represented by polygons, and each polygon has been designated for one of the objectives above. The data[2] comprise two spatial datasets: stand polygons with plot locations, and a tabular dataset consisting of field samples, which are tree measurements. To simplify our presentation, we use a derived dataset that contains fewer attributes.

2.4.6.1 Stand Data

The stand polygon data are stored in an ESRI[3] shapefile, which can be read in using the readShapePoly function of the maptools package.

```
> stands <- readShapePoly("../../data/stands.shp",
+                         verbose=FALSE)
> names(stands)

 [1] "SP_ID"       "AREA"        "PERIMETER"  "STANDID"
 [5] "ALLOCATION"  "TAGE"        "BHAGE"      "DF_SITE"
 [9] "TOTHT"       "CUBVOL_AC"   "TPA"        "QMD"
[13] "BA"
```

The readShapePoly function reads data from a polygon shapefile into a SpatialPolygonsDataFrame object. The data can be accessed using the $ operator (e.g., stands$TAGE) or coerced into a non-spatial data frame using the as.data.frame function. For other geometry types, other functions are available (e.g., ?readShapeLines and ?readShapePoints).

To examine the attribute data for stands, use the names function.

```
> names(stands)

 [1] "SP_ID"       "AREA"        "PERIMETER"  "STANDID"
 [5] "ALLOCATION"  "TAGE"        "BHAGE"      "DF_SITE"
```

---

[2] Available at http://www.cof.orst.edu/cf/forests/mcdonald/.

[3] Environmental Systems Research Institute, Inc.

```
[9]  "TOTHT"      "CUBVOL_AC"  "TPA"            "QMD"
[13] "BA"
```

The names function reveals that the data contain a few geometric attributes (AREA and PERIMETER, in feet), a stand identifier (STANDID), the land classification (ALLOCATION), total age (TAGE), breast-height age (BHAGE), a site productivity measure for Douglas-fir (DF_SITE) (Bruce, 1981), average height of the 40 tallest stems per acre (TOTHT), an estimate of the current volume per acre (CUBVOL_AC), current stocking in trees per acre (TPA), quadratic stand diameter (QMD), and total stand basal area (BA) for each stand polygon. Some of the attributes contain NA (signifying that attributes are missing) and will need to be either ignored or imputed.

To plot the SpatialPolygonsDataFrame object, with unique shadings for the different stand allocations (stands$ALLOCATION), use a gradient in gray scale (?gray) to represent the forest allocation (see Figure 2.8).

2.4.6.2 Plot Data

The plot data, read in using the readShapePoints function,

```
> plots <- readShapePoints("../../data/plots.shp")
```

contains a plot identifier (plots$UNIPLOT) and plot locations stored as coordinates (plots$coords.x1 and plots$coords.x2). The plot locations can be easily added to the existing plot using the plot function with the add = TRUE argument,

```
> plot(plots, add=TRUE, pch=46)

> lev <- as.numeric(stands$ALLOCATION)
> fgs <- gray(length(levels(stands$ALLOCATION)):1 / 3)
> plot(stands,
+       col=fgs[lev],
+       add=FALSE,
+       axes=TRUE)
> title(paste("McDonald-Dunn Research Forest",
+             "Stand Boundaries and Plot Locations",
+             sep = "\n"))
> legend(1280000, 365000,
+        levels(stands$ALLOCATION)[3:1],
+        fill = fgs[3:1],
+        cex = 0.7,
+        title = "Land Allocations")
> plot(plots, add=TRUE, pch=46)
```

McDonald–Dunn Research Forest
Stand Boundaries and Plot Locations

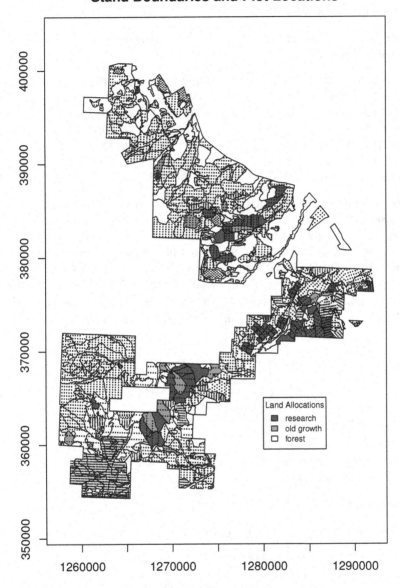

Fig. 2.8: The Oregon State University research forest. The polygons are shaded according to management allocation (`stands$ALLOCATION`), and plot locations are represented as points.

2.4.6.3 Tree Data

Load the tree observation data from the dBase® file using the `read.dbf` function and print a few lines,

```
> mdtrees <- read.dbf("../../data/mdtrees.dbf")
> head(mdtrees)
```

	STANDID	PLOT	TREE	SUBPLOT	SPCODE	DBH	AGE	THT	CBH	SITETREE
1	010101	10001	1	1	WO	15.3	NA	48.7	37.6	0
2	010101	10001	2	1	DF	9.9	NA	60.9	35.0	0
3	010101	10001	3	1	DF	8.1	NA	50.8	26.8	0
4	010101	10001	4	3	GF	4.6	NA	30.3	23.3	0
5	010101	10001	5	3	DF	5.6	NA	46.1	32.4	0
6	010101	10001	6	3	DF	6.5	NA	41.9	25.4	0

To generate plot-level summaries, we first need to compute the per-area expansion factor (acre or hectare weights), denoted π_i, for each tree in the dataset. The sample design is defined by circular plots with radius varying by tree size: 1) a 7.78 ft (2.37 m) radius plot for stems smaller than four inches (10.16 cm) (`SUBPLOT == 3`); 2) a 15.56 ft (4.74 m) radius plot for stems between four and eight inches (20.32 cm) (`SUBPLOT == 2`); and 3) a variable radius plot (20 $\frac{ft^2}{ac}$ BAF) (4.59 $\frac{m^2}{ha}$) BAF) for all stems over eight inches (`SUBPLOT == 1`). Using that definition, the expansion factor for a tree (π_i) is then

$$\pi_i = \begin{cases} \frac{BAF}{B_i} & D > 8.0 \\ \frac{43560.0}{(\pi \times 7.78^2)} & 4.0 < D \le 8.0 \\ \frac{43560.0}{(\pi \times 15.56^2)} & D \le 4.0 \end{cases} \tag{2.1}$$

where B_i is the basal area of tree i and the BAF is 20 square ft per acre.

To compute the expansion factors (π_i), we first create an empty variable in the data frame object.

```
> mdtrees$EXPF <- NA
```

Then we assign the values, using the `SUBPLOT` variable as a filter.

```
> mdtrees$EXPF[mdtrees$SUBPLOT == 1] <-
+      20.0 / (0.0054541539 *
+             mdtrees$DBH[mdtrees$SUBPLOT == 1] ^2)
> mdtrees$EXPF[mdtrees$SUBPLOT == 2] <- 43560 / (pi * 7.78^2)
> mdtrees$EXPF[mdtrees$SUBPLOT == 3] <- 43560 / (pi * 15.56^2)
```

Once again, we print the first few entries of the important columns to verify our results.

```
> head(mdtrees[, 3:11])
```

	TREE	SUBPLOT	SPCODE	DBH	AGE	THT	CBH	SITETREE	EXPF
1	1	1	WO	15.3	NA	48.7	37.6	0	15.66462
2	2	1	DF	9.9	NA	60.9	35.0	0	37.41383
3	3	1	DF	8.1	NA	50.8	26.8	0	55.88980
4	4	3	GF	4.6	NA	30.3	23.3	0	57.26890
5	5	3	DF	5.6	NA	46.1	32.4	0	57.26890
6	6	3	DF	6.5	NA	41.9	25.4	0	57.26890

Now that we have the expansion factors, we can process the tree data and link them to the plot locations, which can yield summaries from which we can create or update stand-level attributes. We can then spatially interpolate attributes between plots and possibly cluster plots into similar sets.

To process the tree data, first `split` the tree `data.frame` object into a `list` object by plot,

```
> trees.by.plot <- split(mdtrees, mdtrees$PLOT)
```

Then, we need to construct a function that

1. sums the expansion factors (π_i) for all trees on each plot (`expf.tot`),
2. sums the expansion factors (π_i) for only those trees with a DBH observation that is not `NA`, which should be all trees over breast height (`expf.ha`),
3. computes the plot's basal area, using only trees over breast height (`ba`),
4. computes the plot's quadratic mean diameter (QMD) (`qmd`),
5. computes the site index using a function from Bruce (1981) (`site`), and
6. returns a vector of the results (`ret.val`),

so that the function can be called using `sapply` to generate a data frame object that will store the results.

Our version of the code for the list of tasks is

```
> get.plot.sums <- function(trs) {
+
+ # /****************************************************/
+ # /* Bruce, D.  1981.  Consistent height-growth and  */
+ # /*     growth-rate estimates for remeasured plots.  */
+ # /*     Forest Science 27:711-725.                   */
+ # /****************************************************/
+     site.index.bruce.1981 <- function(tht, bha) {
+     tht * exp(-21.663 * (3.744e-2 - (bha + 8.0)^ -0.809))
+     }
+
+     not.missing.dbh <- !is.na(trs$DBH)
+     bh.idx <- not.missing.dbh & trs$THT > 4.5
+     expf.tot <- sum(trs$EXPF)
+     expf.bh <- sum(trs$EXPF[not.missing.dbh])
```

```
+     ba <- sum(0.0054541539 * trs$DBH[not.missing.dbh] ^ 2 *
+              trs$EXPF[not.missing.dbh])
+     qmd <- sqrt(ba / expf.bh / 0.0054541539)
+     s.trs <- trs[trs$SITETREE == 1 & trs$SPCODE == "DF" &
+                  !is.na(trs$THT),]
+     nst <- nrow(s.trs)
+     site.bar <-
+         ifelse(nst > 0,
+                weighted.mean(site.index.bruce.1981(s.trs$THT,
+                                                    s.trs$AGE),
+                              s.trs$EXPF),
+              NA)
+     return(c(nrow(trs), expf.bh, expf.tot,
+              ba, qmd, nst, site.bar))
+ }
```

To generate the summaries, call the `sapply` function, and transpose and coerce the results from `sapply` call

```
> plot.sums   <-
+    data.frame(t(sapply(trees.by.plot, get.plot.sums)))
```

so that each row in `plot.sums` contains the plot-level summaries, computed in the `get.plot.sums` function.

The current column names are meaningless, as they have been automatically generated. Also, the plot identifier needs to be appended to our data frame. Therefore we use the names from the `split` operation with the names of the variables from the `get.plot.sums` function,

```
> plot.sums$id <- as.numeric(names(trees.by.plot))
> names(plot.sums) <- c("trees","expf.bh","expf.tot",
+                        "ba","qmd","nst","site","id")
> print(head(plot.sums), digits=3)
```

	trees	expf.bh	expf.tot	ba	qmd	nst	site	id
10001	20	1368.9	2972	134.449	4.244	0	NA	10001
10002	7	1030.8	1260	26.828	2.184	0	NA	10002
10003	2	286.3	286	15.955	3.196	0	NA	10003
10005	4	35.2	722	20.000	10.200	0	NA	10005
10006	3	458.2	687	0.562	0.474	0	NA	10006
10017	8	145.4	145	158.519	14.140	1	117	10017

The next step is to `merge` the plot summary data (`plot.sums`) with the `plot.centers` to create a single data frame object, making sure to preserve empty plots, using the argument `all = TRUE`.

```
> plot.id <- as.numeric(as.character(plots$UNIPLOT))
> plot.centers <- data.frame(cbind(coordinates(plots), plot.id))
```

```
> names(plot.centers) <- c("x","y","id")
> final.plots <- merge(plot.centers, plot.sums, all = TRUE)
> print(head(final.plots[,c(1:3,5:10)]), digits = 3)
```

	id	x	y	expf.bh	expf.tot	ba	qmd	nst	site
1	10001	1264447	400414	1368.9	2972	134.449	4.244	0	NA
2	10002	1264178	400423	1030.8	1260	26.828	2.184	0	NA
3	10003	1263912	400430	286.3	286	15.955	3.196	0	NA
4	10005	1264715	400734	35.2	722	20.000	10.200	0	NA
5	10006	1264460	400741	458.2	687	0.562	0.474	0	NA
6	10017	1265432	398729	145.4	145	158.519	14.140	1	117

We'll use these plots in Chapter 4 so our final step is to write the final plots to a file until we need them,

```
> write.csv( final.plots, "../../data/final-plots.csv")
```

2.4.7 Priest River Experimental Forest

In July and August of 2000, a survey was carried out on the Priest River Experimental Forest, a 2530 ha temperate conifer forest in northern Idaho, USA (48° 212′ N, 116° 472′ W) (Duursma et al., 2003; Pocewicz et al., 2004). Inventory of the forest resources was not the principal goal but a happy by-product.

The forest was divided into nine strata to control the variability that we anticipated would be caused by the topography. The strata comprised three classes of elevation and three classes of solar insolation. Forest growth was expected to vary considerably with elevation because we observed systematic changes in species dominance across elevation gradients, which range from 700 to 1710 m above sea level. Solar insolation is a calculated variable that combines elevation, slope, aspect, and local viewshed to represent the average annual availability of solar radiation at a point. Forest growth was also expected to vary considerably with aspect and slope because the slope variation and the high latitude altered the availability and timing of solar radiation.

Four clusters were assigned to each of the nine strata. The clusters were located randomly within each stratum, subject to the constraint that no point could be more than 500 m from a road. This constraint removed a small portion of the forest from the sampling frame. We will ignore the omission for the purposes of our use of the data.

Each cluster comprised five point locations on a grid. Four of the points were at the corners of a 60 m square, and the fifth was located in the center. At each sample point, a variable-radius plot was installed. The basal area factor was varied adaptively at each site to allow at least six trees per plot. Varying the BAF adaptively in this way leads to a sequential sample, and is generally not recommended because it can introduce bias to the basal area estimates (see, e.g., Banyard, 1987). We will ignore the bias for the purposes of our use of the data.

Each sample tree was measured for species and diameter at 1.3 m (dbh) above mineral soil. A subsample of 85% of the trees was also measured for height. Missing heights were imputed using species-specific diameter–height models constructed from the measured data (Duursma et al., 2003). We will use only those trees with diameter greater than 25.4 cm at 1.3 m.

Tree-level merchantable volumes, in board feet, were estimated using functions provided by Wykoff et al. (1982). The use of the functions requires that diameter will be measured at 1.37 m rather than 1.3 m as in these data. We will ignore the difference for the purposes of our use of the data.

A complete description of the study design can be found in Duursma et al. (2003).

2.4.7.1 Ground Data

These data require a number of processing steps before they can be used for inventory analysis. For example, some of the units are inappropriate, some trees need to be excluded, and some of the labels need to be corrected.

We start by importing the data into R.

```
> pref.tree.all <- read.csv("../../data/pref_trees.csv")
```

After importing the data, we take a snapshot of the data frame and check it for missing values.

```
 [1] "stratum"    "cluster"    "point"      "tree"
 [5] "species"    "distance.m" "azimuth"    "dbh.cm"
 [9] "dbh.in"     "hcb.m"      "ht.m"       "ht.ft"
[13] "baf.ft2.ac" "ba.ft2"     "tf.ac"      "vol.bf"

[1] 1365   16
```

```
> show.cols.with.na(pref.tree.all)
```

```
distance.m    azimuth      hcb.m       ht.m
         2          4        233        235
```

When R imports data from a comma-delimited file, as above, it is forced to guess what class each column is. We should check the class of each variable to make sure that we agree with the guess. We can do so conveniently using sapply. Note that the read.csv command allows the column data classes to be specified using the colClasses argument; see the inbuilt R help files for details.

```
> sapply(pref.tree.all, class)
```

stratum	cluster	point	tree	species
"integer"	"integer"	"integer"	"integer"	"factor"
distance.m	azimuth	dbh.cm	dbh.in	hcb.m
"numeric"	"integer"	"numeric"	"numeric"	"numeric"
ht.m	ht.ft	baf.ft2.ac	ba.ft2	tf.ac
"numeric"	"numeric"	"integer"	"numeric"	"numeric"
vol.bf				
"numeric"				

Everything looks reasonable except the choice of integers for the stratum, cluster, point, and tree identifiers. These variables should all be factors. There is nothing intrinsically wrong with using the integer class here, as R will usually intelligently force the integer to behave like a factor when necessary. For example, this command counts the number of rows within each stratum.

```
> table(pref.tree.all$stratum)

  1   2   3   4   5   6   7   8   9
155 145 158 127 175 155 132 167 151
```

However, making the decision about data type explicit provides us with a layer of error checking. R will warn us if we try to do something that does not make sense, like adding labels together. A one-line for loop takes care of the conversion from integers to factors.

```
> for (i in 1:4)
+   pref.tree.all[,i] <- factor(pref.tree.all[,i])
```

R will then object, correctly, to the illegal operation of trying to average the stratum labels.

```
> mean(pref.tree.all$stratum)

[1] NA
```

A collection of contemporaneous remotely sensed variables is also available for the PREF (Pocewicz et al., 2004). Inevitably, some wrestling is necessary to match data structures when more than one source of data is being used. We document this process because the tools are useful.

The design for the remotely sensed variables is documented more fully below. For the moment, we only need to know that it is based on plots, which correspond to the ground-level clusters. Furthermore, each plot contains nine subplots, which correspond in scale to the points (Pocewicz et al., 2004). The volume measures that comprise our data were made on points that were arrayed in five-point clusters, corresponding to subplots 1, 3, 5, 7, and 9 in the covariate data.

In order to merge the current data frame with the dataset that contains covariates of interest, we need to redefine the point identifier, so that the

current values are mapped from 1, 2, 3, 4, 5, to 1, 3, 5, 7, 9. First, we examine
the current values in the variable.

```
> table(pref.tree.all$point)
```

```
   1    2    3    4    5
 254  304  262  270  275
```

The conversion proceeds as follows.

```
> levels(pref.tree.all$point) <- c("1","3","5","7","9")
```

We then check the new values of the variable.

```
> table(pref.tree.all$point)
```

```
   1    3    5    7    9
 254  304  262  270  275
```

We next consider those trees that are less than 25.4 cm (10 in.) in dbh.
The stand-level volume contribution for trees that are less than 25.4 cm (10
in.) in dbh is negligible. Furthermore, the volume models used to compute
the volume are not reliable for trees that are so small (Wykoff et al., 1982).
We therefore exclude them from the data frame and check the effect on the
sample size as follows.

```
> pref.tree <- subset(pref.tree.all, dbh.cm > 25.4)
> dim(pref.tree)
```

```
[1] 1046   16
```

Note that in removing trees from the database we have removed some rows
from the data frame. In some cases, there may be no rows left that correspond
to certain levels of the factors; that is, the levels may be empty. The factor
levels still remain defined, even though there are no rows corresponding to the
levels. These empty levels can cause problems in later analysis, for example
when looping over levels of the factor. It is therefore best to address the
empty-level condition explicitly. Fixing the empty-level condition is easy to
do: simply redefine the factor.

It is very important, however, to think about which factors to redefine. An
empty level can be informative for a factor that is related to the design. A
plot that has no trees should not be eliminated; to do so would cause bias in
estimates of population parameters. A species that has no trees, on the other
hand, is not of interest. Therefore we redefine the species factor but not the
others. For safety, we surround the relevant command with other commands
that reveal the effect of the redefinition.

```
> levels(pref.tree$species)
```

```
[1] "ABGR"   "ABLA"   "ABLA2" "LAOC"   "PICO"   "PIEN"   "PIMO"
[8] "PIPO"   "PSME"   "THPL"   "TSHE"
```

```
> pref.tree$species <- factor(pref.tree$species)
> levels(pref.tree$species)
```

```
[1] "ABGR"   "ABLA"   "ABLA2" "LAOC"   "PICO"   "PIEN"   "PIMO"
[8] "PIPO"   "PSME"   "THPL"   "TSHE"
```

In the end, no species levels needed to be dropped.

The tree-level volume was computed in board feet, which is a unit of volume equivalent to 144 cubic inches, but when volume in board feet is computed, deductions are made for squaring assumptions and kerf. That is, in general, in the systems for which these functions were originally constructed, if volume is reported in cubic feet, then it reflects the physical volume from which timber may be cut, whereas if volume is reported in board feet, then it reflects the physical volume of the bole discounted for the loss of volume during the cutting process.

We will convert this volume measure from board feet to cubic meters. Also, the tree factor must be converted from trees per acre to trees per hectare. Note that the tree factor is the sampling weight for each tree, not the factor that identifies the tree identity!

```
> pref.tree$vol.m3 <- pref.tree$vol.bf / 12 * 0.0283168466
> pref.tree$tf.ha <- pref.tree$tf.ac * 2.47105381
> pref.tree$vol.m3.ha <- pref.tree$vol.m3 * pref.tree$tf.ha
> pref.tree$baf.m2.ha <-
+    pref.tree$baf.ft2.ac / 3.2808399^2 / 0.404685642
```

We finally need to add the sampling weight to the tree database. This is somewhat easier than trying to add the sampling probability. For the PREF inventory, the weight is theoretically rather complex because of the sequential nature of the survey; however, as we noted above, we are ignoring that element. Therefore, the sampling weight for each tree is proportional to its basal area and inversely proportional to the basal area factor of the wedge used for its plot, so it is inversely proportional to the tree factor.

```
> pref.tree$weight <- 1/pref.tree$baf.ft2.ac
```

The tree-level volume per hectare is the product of the tree volume in cubic meters and the tree factor in trees per hectare. This quantity can then be summed to the point level. There are numerous ways to make that sum, of which aggregate is the most convenient because it creates a complete data frame.

```
> pref.point <- with(pref.tree,
+                     aggregate(x = list(ba.m2.ha = baf.m2.ha,
+                                        vol.m3.ha = vol.m3.ha),
```

```
+                              by = list(stratum = stratum,
+                                        cluster = cluster,
+                                        point = point),
+                            FUN = sum))
```

We should examine the new object to ensure that it matches our expectations. For example, we expect each of the nine strata to contain 20 points.

```
> table(pref.point$stratum)

 1  2  3  4  5  6  7  8  9
18 14 20 16 19 20 20 20 18
```

This summary shows a problem: several strata are missing points. The aggregate function drops indicators for the empty points. At the time of writing, there is no way to instruct aggregate to keep the empty levels, although there may well be in the future. Regardless, we can take the opportunity to demonstrate more data manipulation commands. As always, there are several options for proceeding.

We will create a new data frame that reflects the structure of the design, using the expand.grid function, and merge it with this dataset. Note that the cluster identifier crosses stratum boundaries, instead of being nested inside the strata, so we will add the stratum identifier to our new data frame after creating it.

```
> design.point <-
+   expand.grid(cluster = levels(pref.tree.all$cluster),
+               point = levels(pref.tree.all$point))
> str(design.point)
```

```
'data.frame':           180 obs. of  2 variables:
 $ cluster: Factor w/ 36 levels "1","4","5","6",..: 1 2 ...
 $ point  : Factor w/ 5 levels "1","3","5","7",..: 1 1 ...
 - attr(*, "out.attrs")=List of 2
  ..$ dim      : Named int  36 5
  .. ..- attr(*, "names")= chr  "cluster" ...
  ..$ dimnames:List of 2
  .. ..$ cluster: chr  "cluster=1" ...
  .. ..$ point  : chr  "point=1" ...
```

This new data frame comprises 36 plots of five clusters for a total length of 180. Next we need to construct a data frame that contains the cluster and the stratum identifiers for all the clusters, including the empty ones. We could create such a document in a spreadsheet and read it in. We prefer to reduce the number of steps needed to analyze the data by creating the needed object in R. Here, we do this by extracting the cluster and stratum columns from the original dataset, removing all the missing values, and then removing all the duplicate rows, as follows.

```
> design.cluster <-
+    unique(pref.tree.all[, c("cluster","stratum")])

> str(design.cluster)

'data.frame':        36 obs. of  2 variables:
 $ cluster: Factor w/ 36 levels "1","4","5","6",..: 1 7 ...
 $ stratum: Factor w/ 9 levels "1","2","3","4",..: 1 2 ...
```

The merging operation requires the merge command. This operation is best monitored, and an easy way to monitor it is to use dim commands before and afterward. When reusing an object name, we should check the effect of our merge command on a test data frame first, so we do not have to recreate the original if we make a mistake. And, we often do. Also, it is important to ensure that the factors for merging are appropriately defined before merging. Failure to do so can lead to errors that are difficult to detect, and may only show up much later in the analysis or, worse, go undetected.

```
> dim(design.point)

[1] 180    2

> dim(design.cluster)

[1] 36   2

> test <- merge(x = design.point,
+               y = design.cluster,
+               all = TRUE)
> dim(test)

[1] 180    3

> head(test)

  cluster point stratum
1       1     1       1
2       1     7       1
3       1     9       1
4       1     5       1
5       1     3       1
6      10     1       2
```

This seems to have worked just fine. We can now merge the volumes.

```
> design.point <- test
> dim(design.point)

[1] 180    3

> dim(pref.point)
```

```
[1] 165    5

> test <- merge(x = design.point,
+               y = pref.point,
+               all = TRUE)
> dim(test)

[1] 180    5

> head(test)

  cluster point stratum    ba.m2.ha    vol.m3.ha
1       1     1       1     1.147842     4.284489
2       1     3       1    22.956841    86.528810
3       1     5       1     9.182736    42.078591
4       1     7       1    13.774105    67.786741
5       1     9       1    12.626263    65.055821
6       4     1       1    45.913682   164.223243
```

Again, this is successful. Three last steps complete the cleaning.

```
> test$vol.m3.ha[is.na(test$vol.m3.ha)] <- 0
> test$ba.m2.ha[is.na(test$ba.m2.ha)] <- 0
> pref.point <- test
> rm(test)
```

Our final step is to append the survey weights. In this case, the weights are equal, which simplifies the process.

```
> pref.point$weight <- 1
```

This seems like a lot of work to complete a conceptually very simple operation. It simultaneously shows a weakness and some strengths of R: sometimes what you want to do takes numerous small steps, but the advantage is that the decisions that you have made along the way are explicitly documented, and the scripts are reusable.

2.4.7.2 Remotely Sensed Data

A Landsat 7 ETM+ image of the forest was procured at the same time as the field measurements were under way (Pocewicz et al., 2004). Briefly, the Landsat image covered the entire forest and comprised square pixels of approximately 30 m side length. Several popular vegetation indices were computed. We choose the Normalized Difference Vegetation Index with a correction for the middle-infrared wavelength, which showed the best performance for predicting an effective plant area index (Pocewicz et al., 2004). We assume that effective plant area is correlated with above-ground merchantable

tree volume. Full information on the processing steps applied can be found
in Pocewicz et al. (2004) and Landsat Project Science Office (2005).

We now access two relevant comma-delimited datasets. The pixel dataset
contains the corrected, pixel-level response values for the coverage of the
PREF.[4] The subplot dataset contains field and remote measurements for the
points. The remotely sensed covariates are taken from the pixels in the pixel
database that were closest in location to the field points.

```
> pref.pixel <- read.csv("../../data/pref_pixels.csv")
> names(pref.pixel)

 [1] "num"      "utme"      "utmn"        "lai.rs"
 [5] "ndvi"     "mir"       "slope"       "cti"
 [9] "elev"     "inso"      "ndvic"       "mirc.1"
[13] "mirc.2"   "mirc.amy"  "mirc.check"  "mirc.pref"

> dim(pref.pixel)

[1] 28102    16

> pref.pixel[1:5, c(1:5,11)]

  num    utme    utmn   lai.rs      ndvi     ndvic
1   1 510731.2 5353596 5.870416 0.8032642 0.7116639
2   2 510731.2 5353626 5.229210 0.7827569 0.6076666
3   3 510731.2 5353656 5.900865 0.7757183 0.6872592
4   4 510731.2 5353566 4.912140 0.7921855 0.6149861
5   5 510731.2 5353686 6.003587 0.7757183 0.6872592

> show.cols.with.na(pref.pixel)

No missing values.

> sapply(pref.pixel, class)

       num        utme       utmn      lai.rs       ndvi
 "integer"   "numeric"  "numeric"   "numeric"  "numeric"
       mir       slope        cti        elev       inso
 "integer"   "numeric"  "numeric"   "numeric"  "integer"
     ndvic      mirc.1     mirc.2    mirc.amy mirc.check
 "numeric"   "numeric"  "numeric"   "numeric"  "numeric"
 mirc.pref
 "numeric"
```

The pixel-level dataset seems fine.

```
> pref.subplot <- read.csv("../../data/pref_subplots.csv")
> names(pref.subplot)
```

---

[4] Landsat imagery courtesy of NASA Goddard Space Flight Center and U.S. Geo-
logical Survey.

```
[1]  "plot"            "subplot"          "point"
[4]  "scale"           "utme"             "utmn"
[7]  "slope"           "wetind"           "elevclass"
[10] "insoclass"       "elev1"            "elevsq1"
[13] "inso1"           "insosq1"          "LAIbeer2total"
[16] "LAIbeer2can"     "acblue"           "acgreen"
[19] "acred"           "acnir"            "acmir1"
[22] "acmir2"          "acndvi"           "acndviC"
[25] "acsr"            "acsrC"            "acvegind"
[28] "acsavi"          "acsarvi2"         "acslavi"
[31] "acinfrared"      "acbrightness"     "acgreenness"
[34] "acwetness"       "MIRc"
```

```
> dim(pref.subplot)
```

```
[1] 324  35
```

```
> pref.subplot[1:5, c(1:3, 5:6, 24)]
```

```
  plot subplot point   utme     utmn    acndviC
1    1       1     1 a 515397 5357562 0.3328538
2    1       2     2 a 515397 5357592 0.7114403
3    1       3     3 a 515397 5357622 0.7114403
4    1       4     4 a 515427 5357622 0.7396296
5    1       5     5 a 515457 5357622 0.6101370
```

```
> show.cols.with.na(pref.subplot)
```

```
No missing values.
```

```
> sapply(pref.subplot, class)[1:4]
```

```
     plot    subplot      point      scale
"integer"  "integer"   "factor"   "factor"
```

```
> pref.subplot$plot <- factor(pref.subplot$plot)
> pref.subplot$subplot <- factor(pref.subplot$subplot)
```

The subplot-level dataset also seems fine. The merge command blends our existing point-level information with the new subplot-level information. Again, we surround our merge with dim statements to monitor the process.

```
> dim(pref.point)
```

```
[1] 180   6
```

```
> dim(pref.subplot[,c("plot","subplot","acndviC")])
```

```
[1] 324   3
```

```
> pref.point.cov <-
+   merge(x = pref.point,
+         y = pref.subplot[,c("plot","subplot","acndviC")],
+         all.x = TRUE, all.y = FALSE,
+         by.x = c("cluster","point"),
+         by.y = c("plot","subplot"))
> head(pref.point.cov)
```

```
  cluster point stratum  ba.m2.ha vol.m3.ha weight   acndviC
1       1     1       1  1.147842  4.284489      1 0.3328538
2       1     3       1 22.956841 86.528810      1 0.7114403
3       1     5       1  9.182736 42.078591      1 0.6101370
4       1     7       1 13.774105 67.786741      1 0.6126499
5       1     9       1 12.626263 65.055821      1 0.8090011
6      10     1       2  4.591368 16.293566      1 0.5550567
```

```
> dim(pref.point.cov)
```

```
[1] 180    7
```

```
> show.cols.with.na(pref.point.cov)
```

```
No missing values.
```

We now have a point-level dataset with the necessary design information, a variable of interest, and an auxiliary variable.

We note that there is a mismatch between the scales of the two sets of measurements. The Landsat data comprise the average spectral responses across 30 meter pixels. The ground measurements represent points on the landscape, ideally in the center of the pixel but most likely not. There also may be errors of registration; that is, the points and the pixels might not line up as we would hope. These types of errors are likely to occur in large-scale inventories. We will deal with this mismatch by defining the sampling unit as the 30 meter area represented by each Landsat pixel. The point-level volume measurements are then a subsample of the volume per hectare within the pixel. Understood this way, the design is a two-stage–two-phase sample, with auxiliary information on the primary sampling unit (PSU) and the variable of interest on the secondary sampling unit (SSU), of which there is only one per PSU. Särndal et al. (1992) devote a section to this problem. We will ignore it because the within-PSU variation is impossible to assess with only one SSU, and we believe that it is likely to make only a negligible contribution anyway (see Section 3.4.2.1).

2.4.8 Leuschner

For our forest-planning example, we have used the data presented in Leuschner (1990, see Tables 3.1 and 3.2) and Curtis et al. (1982). The data, originally from DFSIM simulations, was adapted and used to demonstrate forest planning and regulation techniques by Leuschner (1990; see Tables 3.1 and 3.2).

For this book, we have entered the values into a text file that can be read into R using the `read.table` function

```
> leusch.ylds <- read.table("../../data/leuschner.txt",
+                          header = TRUE)
```

where the resulting `leusch.ylds` object contains four columns: 1) the stand identifier (`stand`), 2) the planning period (`per`), 3) the age of the stand in period `per`, and 4) the yield (*volume*), in thousands of cubic feet per acre, if we harvest stand `stand` in period `per`.

In these data, there are eight stands and the problem is to schedule the harvest of the forest over six periods to create a fully regulated forest. The areas of the stands are presented in Chapter 9, where we use these data to demonstrate using R for forest activity scheduling.

2.5 Summary

We have briefly covered many of the basic data management and manipulation functions within the context of data conversion, error checking, generating summaries and plots. The tools that R contains for data manipulation (keyword 'manip') are numerous, and we have only scratched the surface. Some of the documentation is extensively detailed, and reading it may seem laborious. It is well worth the effort.

In the next chapter, we will start processing data from forest samples and discuss some sampling design topics.

Part II
Sampling and Mapping

Chapter 3
Data Analysis for Common Inventory Methods

3.1 Introduction

This chapter covers the analysis of sample surveys in the context of natural resources inventories, with a particular focus on forest inventories. Our goal is twofold: to provide coverage that is relevant to the efficient analysis of sample survey data using R, and to demonstrate the data manipulation techniques introduced in Chapter 2. We also provide scripts and commentary for the analysis of sampling designs that are commonly used in natural resource inventories.

We include only as much background material as necessary to motivate our analysis. More thorough expositions on various elements of sampling theory are available elsewhere (see, e.g., Cochran, 1977; Särndal et al., 1992; Schreuder et al., 1993). Also, our survey of inventory designs is limited by space and variation. We focus on those designs that permit us to present the key elements of analysis in R rather than trying to be exhaustive in scope.

We draw heavily on several third-party packages. Notably, Thomas Lumley's survey package provides most of the necessary sampling estimation tools (Lumley, 2004, 2010), and Angelo Canty's boot package, maintained by Brian Ripley, provides useful tools for bootstrapping (Davison and Hinkley, 1997; Canty and Ripley, 2010). We include our own R code where it seems useful. The code we present is not necessarily optimal for our immediate purpose; there are many ways to proceed using R, and the palette of options increases every month. We intend that the data-processing sections serve two purposes: first, to demonstrate how to achieve simple but important goals in the processing of natural resources inventory data, and second, to demonstrate some of the flexibility and power of the R language.

We next introduce some useful vocabulary and examples for sample surveys in the context of forest inventories. We then identify three example datasets that we will use to demonstrate the steps necessary for point and interval estimation from sample surveys. In Section 3.2, we introduce interval computation and demonstrate three methods for their estimation. The following sections detail the use of R for single-level sampling (Section 3.3), hierarchical sampling (Section 3.4), and sampling with auxiliary information (Section 3.5).

3.1.1 Infrastructure

We briefly introduce some useful vocabulary and examples for sample surveys in the context of forest inventories.

The *population* is the entity for which one wishes to estimate quantities of interest. The population comprises sampling units and is in a sense co-defined with the sampling unit. It must be possible to represent the population by a *frame*. An example of a population is the Priest River Experimental Forest (PREF).

A *frame* is a list of units of the population that is established to facilitate the selection of a random sample from the population. A frame for the PREF might be a map with a grid drawn upon it.

The *sampling unit* is the unit that we select from the population to make measurements upon. A sampling unit for the PREF might be a fixed area within the forest of size, say, 0.1 ha.

The *sample* is the collection of sampling units that are measured.

The *variables of interest* are the quantities that are measured on the sampling units about which we wish to make estimates.

A *statistic* is an arithmetical function of data that reduces it to a summary. Example statistics are the total, the mean, and the standard deviation.

The *parameters* are the population-level statistics of the variables of interest.

The *sampling distribution* is the distribution of values that could possibly be taken by a statistic that is computed using a sample of data taken from a population.

Sometimes we will also measure *auxiliary information*, which is information that is in some way related to the variable of interest and is available for all or most of the units in the population.

For example, we may wish to estimate the population total (a parameter) of the volume of merchantable timber (variable of interest) on PREF. We take a map of the PREF, divided into a 0.1 ha grid (the frame), and randomly nominate a collection of 100 grid spaces (plots) to locate and measure (these plots are the sample). Upon each plot we measure the volume of the trees (the variable of interest). We may have soil, elevation, and aspect measures for the entire forest (auxiliary information). We can estimate the population total using a statistic that is calculated from the variables of interest as measured on the sample. We will use an assumption about the sampling distribution

of the sample statistic to estimate the confidence interval of the estimate of
the population parameter.

3.1.2 Example Datasets

We adopt three datasets for our examples. In each case, we report the sample
design that was used without necessarily making any recommendations as to
the appropriateness of the design for other purposes. In order to use these
example datasets in this chapter, it was necessary to simplify or ignore elements of the actual design. We will document the necessary simplifications
when they are applied. The datasets are the UFC data (Section 2.4.1), the
PREF data (Section 2.4.7), and the FIA data (Section 2.4.3).

3.2 Estimate Computation

An estimate of a population parameter is more useful when it is accompanied
by an estimate of its certainty. Two measures of certainty that are commonly
computed are the standard error and the confidence interval.

Confidence intervals can be computed in several ways. Here we will cover
classical large-sample theory, jackknife, and bootstrap methods. We review
the principles of the sampling distribution in order to set the scene for development of the interval estimates.

3.2.1 Sampling Distribution

Given an assumed sampling distribution of a statistic, our goal is to identify
an interval of that distribution that contains a suitable proportion of the
sampling distribution, for example 95%. We need to estimate the shape of
the distribution to be able to identify the interval that contains a suitable
proportion. Each of the three strategies that we demonstrate for computing
the interval uses a different approach to solving this problem.

For reference, the confidence interval for the population mean can be expressed as

$$\hat{\mu} \pm \hat{s}_{\hat{\mu}} \times t^{n-1}_{1-\alpha/2} \tag{3.1}$$

where $\hat{\mu}$ is an estimate of the population mean (often the sample mean), $\hat{s}_{\hat{\mu}}$ is
an estimate of the standard error of the estimate of the population mean, and

$t^{n-1}_{1-\alpha/2}$ is a quantile from Student's t-distribution. The following approaches can be applied.

1. Large-sample theory invokes the Central Limit Theorem and Slutsky's Theorem to permit the assumption that the sampling distribution of the mean is Student's t. The quantiles of Student's t are used with the sample-based estimate of standard error $\hat{s}_{\hat{\mu}}$ to compute a confidence interval, as per equation (3.1). See Section 3.2.2.
2. When no reasonable closed-form estimate for the standard error can be found (e.g., for the ratio of means estimator) approximations may be used. A popular example is the Taylor-series linearization, which expands the function of the estimators into an infinite series and ignores all terms beyond the linear approximation (see, e.g., Särndal et al., 1992, p. 172). We cover linearization in Section 3.2.3.
3. Use of the jackknife allows us to replace the analytical estimate of the standard error $\hat{s}_{\hat{\mu}}$ in equation (3.1) with a resampling-based estimate. This approach is useful when no reasonable closed-form estimate for the standard error can be found (e.g., for the ratio of means estimator). Use of the jackknife requires the assumption that the sampling distribution be normal, or at the very least symmetric. See Section 3.2.4.
4. The bootstrap generates an estimate of the shape of the sampling distribution directly by resampling from the observed sample, and computing the statistic of interest, many times. See Section 3.2.5. The bootstrap can be used in one of two ways, each of which has numerous variations.

   a. Bootstrap-based estimates of the standard error $\hat{s}_{\hat{\mu}}$ can be used to replace the usual sample-based estimate in equation (3.1) when no reasonable closed-form estimate of the standard error is known (e.g., for the ratio of means estimator).
   b. Bootstrap-based estimates of the confidence interval can be used instead of the usual formula (3.1) if there is doubt about the shape of the sampling distribution.

Among these options, if there is a closed-form equation for the standard error of the parameter of interest and the sampling distribution of the parameter of interest can be reasonably expected to be normal (or at least symmetric), then approach 1 is reasonable. If there is no closed-form equation for the standard error of the parameter of interest but the sampling distribution of the parameter of interest can be reasonably expected to be normal (or at least symmetric), then estimating the standard error $\hat{s}_{\hat{\mu}}$ using approach 3 or 4a is reasonable. If there is doubt about the shape of the sampling distribution, then approach 4b is best, but careful use of 4a is still possible (see Section 3.2.5.2).

There is no definitive answer as to whetherthe jackknife or the bootstrap is better for estimating the standard error of an estimate. Schreuder et al. (1993) summarized a collection of studies that used forestry data and addressed common forestry concerns, and their conclusions depended heavily on the sample design.

In practical terms, the jackknife requires considerably less computation time than the bootstrap, and is probably computationally more stable, because the jackknife resamples are most likely more similar to the original sample than are the bootstrap samples. All else being equal, if an estimator can be computed for a sample then it is more likley that it can also be computed for a jackknife sample than for a bootstrap sample. These observations suggest that, unless there is a strong reason to do otherwise, using the jackknife is a reasonable strategy.

Several books have been written that compare the traditional, jackknife, and bootstrap estimates for various purposes using simulations and asymptotic theory. For example, Shao and Tu (1995) develop and compare the rates of convergence of various estimators and report simulation studies. The authors focus on sample surveys, as well as more general applications, and demonstrate that for sample surveys the jackknife and the linearization estimators have the best performance (Shao and Tu, 1995, pp. 252–256). Manly (1997) compared the coverage rates and accuracy from a simulation study for estimating a confidence interval for the standard deviation of a sample of size 20 generated from an exponentially distributed population. The small-sample properties of the compared estimators performed poorly, but such a case seems more extreme than those typically faced in natural resources inventories, for which larger samples and relatively straightforward functions of the mean will be more common.

Overall, the jackknife estimator has no asymptotic advantage over the linearization estimator (Shao and Tu, 1995, p. 262). Lumley noted that the linearization estimates "...are approximately unbiased but may be quite variable and as a result tend to lead to confidence intervals that are too short in small samples" (Lumley, 2004, p. 4). However, Shao and Tu (1995) present simulation results that suggest that the linearization estimator can be more (negatively) biased and less variable than the jackknife estimator (Table 2.1, p.32) in small samples.

Other approximations are plausible, and may become more popular. Saddlepoint approximations are more complicated functions of the sample statistics that provide more accurate approximations to the distribution and therefore are more likely to be accurate at a given sample size. Readable expositions can be found in Reid (1988) and Goutis and Casella (1999). A forest inventory application is reported in Magnussen (2001) and an examination of the use of the approximation in ratio and regression estimation in Agho et al. (2005).

Higher-order asymptotics also provide opportunities for improvement (see, e.g., Reid and Fraser, 2000; Reid, 2003). The "higher-order" label refers to the inclusion of terms in the approximations that are higher than linear order, which results in a faster convergence rate of the estimators. Again, the result is to provide estimators with better asymptotic properties than the existing estimators; for example, more accurate intervals. Higher-order asymptotics are implemented in R for some models (Brazzale, 2005).

3.2.2 Intervals from Large-Sample Theory

We ordinarily use large-sample theory, such as the Central Limit Theorem and Slutsky's Theorem (see, e.g., Casella and Berger, 1990), to tell us how we can reasonably treat the sampling distribution of the parameter of interest. This leads to the use of the familiar estimators in, among others, Bell and Dillworth (1997), Avery and Burkhart (2003), and Iles (2003).

For example, the sample mean is used to estimate the population mean in simple random sampling. The sampling distribution of a mean can be reasonably treated as though it were normal if the sample is large enough. (The population total is more commonly the parameter of interest in natural resources inventories, but the mean provides a more straightforward example.) The best estimate of the population mean is the sample mean, as it is unbiased and has the least variance among all unbiased estimators.

$$\hat{\mu}_v = \bar{v} = \frac{1}{n}\sum_{i=1}^{n} v_i \tag{3.2}$$

The standard error of the mean is estimated by

$$s_{\bar{v}} = \frac{s_v}{\sqrt{n}} \tag{3.3}$$

where s_v is the standard deviation of the sample. (In the finite population case, we will add the finite population correction; see equation (3.3).)

Then an approximate 95% confidence interval for the population mean is

$$95\% \text{ C.I.} = \bar{v} \pm s_{\bar{v}} \times t_{0.975,n-1} \tag{3.4}$$

These quantities can be easily computed in R (see Section 3.3.1.1).

The utility of these theories in any given situation depends on the circumstances. Generally, as a prescription, we invoke the theories if the sample size is large enough. What that prescription means in practice is anyone's guess.[1]

It is straightforward to generalize the approach above to cases where the population parameter of interest is a linear function of other population pa-

---

[1] The Berry-Esseen Theorem, which describes the rate of convergence of the sampling distribution to normality, states that the maximum vertical distance of the cdf of the standardized sample mean of data with finite third moment to the normal cdf is constrained as a function of a constant, the sample size, and the skew of the data, and sigma raised to the power 3 (Berry, 1941; Esseen, 1942). Hence, the skew of the population distribution of the data affects the quality of the approximation, but the kurtosis does not.

rameters, for example, in stratified sampling. The means and variances of independent random variables are additive.

3.2.3 Intervals from Linearization

Use of the approach developed in Section 3.2.2 is impossible when the population parameter of interest is estimated by a non-linear function of random variables. An example is the *ratio of means* estimate for ratio estimation, for which the estimate is the ratio of two sample means, each of which is a random variable. The usual theory to compute linear functions of random variables will not help us here; we need another approach.

Linearization involves finding a close linear approximation to the non-linear function. For example, one could approximate the estimator by a first-order Taylor-series expansion, for which it is relatively easy to estimate the variance. Thus, we replace the exact function, which has an unknown variance, by a close approximation, which has a variance that is easily calculated. When the approximation is linear, then the approach is called "first order". The linearization approximation is also known as the delta method, and is related to the sandwich estimators that are used in some branches of statistics.

An example of a linearized estimate is the ratio of means estimate for ratio estimation, which we briefly cover here. We base our derivation on Wolter (1985). We have a variable of interest, y_i, measured on a simple random sample of n units from an infinite population. We also have an auxiliary variable, x_i, measured on the same units, and we know the population mean of x, which we denote μ_x. We wish to estimate μ_y. For the ratio of means estimator,

$$\hat{\mu}_y = \mu_x \times \hat{R} = \mu_x \times \frac{\bar{y}}{\bar{x}} \tag{3.5}$$

Since we know μ_x, we can write the important parts of equation (3.5) as a function of a bivariate parameter θ: $\hat{R} = g(\theta)$, where $\theta = (\bar{x}, \bar{y})$ and $g(\theta) = \frac{\theta_2}{\theta_1}$. The linearized estimate of the variance is then based on the first-order Taylor-series expansion of the estimator:

$$s_R^2 \simeq \mathbf{d} \Sigma \mathbf{d}' \tag{3.6}$$

where $\mathbf{d}$ is a vector of first derivatives of g with respect to the elements of θ and Σ is the covariance matrix of θ[2]. Then,

---

[2] Wolter (1985) notes that equation (3.6) is the estimate of the MSE of the estimator, rather than the variance, but that the bias is negligible to a first-order approximation.

$$\mathbf{d} = \begin{bmatrix} \frac{\delta\theta_1}{\delta g} & \frac{\delta\theta_2}{\delta g} \end{bmatrix}$$

$$= \begin{bmatrix} \frac{\delta\bar{x}}{\delta g} & \frac{\delta\bar{y}}{\delta g} \end{bmatrix}$$

$$= \begin{bmatrix} -\frac{\bar{y}}{\bar{x}^2} & \frac{1}{\bar{x}} \end{bmatrix} \tag{3.7}$$

and

$$\Sigma = \begin{bmatrix} s_{\bar{x}}^2 & s_{\bar{x}\bar{y}} \\ s_{\bar{x}\bar{y}} & s_{\bar{y}}^2 \end{bmatrix} \tag{3.8}$$

Straightforward algebra shows us that

$$\hat{s}_{\hat{R}}^2 = \mathbf{d}\Sigma\mathbf{d}' = \frac{s_{\bar{x}}^2 \bar{y}^2}{\bar{x}^4} - \frac{2s_{\bar{x}\bar{y}}\bar{y}}{\bar{x}^3} + \frac{s_{\bar{y}}^2}{\bar{x}^2}$$

$$= \frac{R^2}{n}\left[\frac{s_y^2}{\bar{y}^2} + \frac{s_x^2}{\bar{x}^2} - 2\frac{s_{xy}}{\bar{x}\bar{y}} \right] \tag{3.9}$$

Equation (3.9) is equivalent to those found in Cochran (1977), Wolter (1985), and Schreuder et al. (1993). The estimate of the variance of the estimate of the population mean is then

$$s_{\hat{\mu}_y}^2 = \mu_x^2 \times \hat{s}_{\hat{R}}^2$$

$$= \frac{\mu_x^2}{\bar{x}^2} \times \frac{s_y^2 + \hat{R}^2 s_x^2 - 2\hat{R}s_{xy}}{n} \tag{3.10}$$

which is similar to that in Avery and Burkhart (2003) and identical if $\bar{x} = \mu_x$ and the finite population correction is considered.

Linearization-based estimates of standard errors are provided by the survey package of R (Lumley, 2004).

3.2.4 Intervals from the Jackknife

The application of the jackknife that is relevant to our interests is to provide estimates of standard errors for non-linear functions of random variables. There are references to two types of jackknife, although one is obviously a special case of the other: the delete-1 jackknife and the delete-d jackknife. We consider only the first, and refer to it hereafter as the jackknife.

The principle behind the jackknife is to divide the sample into subsamples and compute a statistic for each subsample. Then each such statistic provides information about the distribution of the parameter of interest. The following brief description draws from Shao and Tu (1995).

Assume that g is a possibly biased estimator of a parameter of interest θ, where g is a function of the data from a sample of size n. Let g_{-i} denote the same function, computed on the same sample, but with the i-th observation removed and the sample size reduced to $n-1$. Then g and each of the g_{-i} are all valid, possibly biased, estimators of θ.

Now define each of n pseudovalues as follows. The i-th pseudovalue, $\tilde{g}_i$, is

$$\tilde{g}_i = n \times g - (n-1) \times g_{-i} \qquad (3.11)$$

Each pseudovalue is now an estimate of θ. Also note that each pseudovalue is, loosely speaking, the difference between the function g applied to two related samples. Therefore a portion of the bias in g will be canceled out in the $\tilde{g}_i$ and any estimates that are calculated from them.

Tukey (1958) asserted that these pseudovalues could be treated as though they were independent and identically distributed. Furthermore, each pseudovalue has a variance approximately equal to $n \times \sigma_g^2$, where σ_g^2 is the variance of g. If this is true, then we can estimate θ by the mean of the pseudovalues, and the standard error of the estimate of θ will be the standard error of that mean. That is,

$$\hat{\theta}_{jack} = \frac{1}{n} \sum_{i=1}^{n} \tilde{g}_i \qquad (3.12)$$

$$\hat{s}^2_{\theta_{jack}} = \frac{1}{n(n-1)} \sum_{i=1}^{n} \left(\tilde{g}_i - \hat{\theta}_{jack}\right)^2 \qquad (3.13)$$

This estimator $\hat{\theta}_{jack}$ will have less bias than g, and the estimate of the standard error can be computed easily for arbitrary functions.

Jackknife-based estimates of standard errors are provided by the survey package of R (Lumley, 2004). In cluster and multi-stage designs the jackknife variance is computed by jackknifing the highest-stage units only.

3.2.4.1 A Brief History of the Jackknife

The origin of the jackknife is curiously unsettled. There seems little doubt that it was suggested originally by Maurice Quenouille as a tool for bias reduction. Casella and Berger (1990, p. 341) and Cochran (1977, p. 178) refer to Quenouille (1956), as does Tukey (1958). Efron and Tibshirani (1993, p. 133) say that Quenouille proposed it in the mid-1950s but later cite Quenouille (1949a), as do Davison and Hinkley (1997, p. 59) and Shao and Tu (1995, p. 4). Särndal et al. (1992, p. 437) and Schreuder et al. (1993, p. 102) both cite Quenouille (1949b) and Quenouille (1956). Quenouille (1956, p. 358) does cite Quenouille (1949a) as referring to a special case of the principle. To us, Quenouille (1956) seems to be the clearest starting point.

Most authors then cite Tukey (1958) as providing the original suggestion for variance estimation, although the citation is only to an abstract of a preliminary report presented at an April 1958 meeting of the Institute of Mathematical Statistics. Tukey cites Quenouille (1956) as suggesting the use of the tool to reduce bias, and Jones (1956) for the idea of using subsamples of the sample in general.

To round out our historical exegesis, the earliest explicit reference to the name "jackknife" that we can find in a peer-reviewed publication is either Brillinger (1964) or Miller (1964). David (1995) cites Miller (1964) as being the earliest peer-reviewed reference. We note that both articles were in the last edition for 1964 of each journal, Miller's being in December, and that Miller cites Brillinger, but as being "soon to appear". Concerning the question about lineage above, Brillinger cites Quenouille (1956), Tukey (1958), and a 1959 unpublished manuscript by Tukey,[3] and Miller cites Quenouille (1949a) and a 1962 unpublished manuscript by Tukey.[4] Only the latter seems to be among the published collection of Tukey's papers (Jones, 1986).

3.2.5 Intervals from the Bootstrap

The history of the bootstrap is much easier to trace than that for the jackknife: it was introduced by Efron (1979). We focus here on the non-parametric bootstrap.

The bootstrap can be used to obtain an estimate of the empirical sampling distribution of many useful statistics and functions of statistics. The estimate of the empirical sampling distribution can be examined directly for quantities of interest, such as standard errors, or confidence intervals of parameters of interest.

An important caveat is that the bootstrap can ameliorate some of the problems associated with inference from small samples but will not eliminate them. The bootstrap relies upon its own large-sample theory, and it has its own strengths and weaknesses. Published theoretical examinations suggest that in specific situations the convergence of the bootstrap large-sample theory is faster than in the classical large-sample theory (Hall, 1992; Shao and Tu, 1995). Consequently, it is important to recognize that clever use of the bootstrap can lead to estimators with significantly better properties than naive use of the bootstrap.

A good, basic introduction to and description of the bootstrap can be found in Efron and Tibshirani (1993). For further reading, we suggest Davison and Hinkley (1997) and Manly (1997). More advanced expositions can be found in Hall (1992) and Shao and Tu (1995).

---

[3] Tukey, J. 1959. Approximate confidence limits for most estimates. Unpublished manuscript.

[4] Tukey, J. 1962. Data analysis and behavioral science. Unpublished manuscript.

3.2.5.1 Implementation

We start with a simple situation: computing a bootstrap estimate of the standard error of the sample mean in R. We will obtain bootstrap estimates of the standard error of the mean for the purposes of demonstration only; the usual formula (3.3) is perfectly satisfactory. However, if the goal is an interval estimate, then the assumption of normality of the sampling distribution should be carefully considered.

First, we need a function that computes the mean of the sample. The function must accept an index and use that index to permute the sample data, as follows.

```
> boot.mean <- function(x, index) {
+    mean(x[index])
+ }
```

Superficially this function seems like it must be more complicated than is necessary, and, for this problem, perhaps it is. However, for more complicated situations, this syntax lends itself to very easy generalization. The `boot` function will efficiently call this function many times, sending the same data each time, along with a randomly generated index, which permutes the sample. We demonstrate the function using only the first point in each cluster of the PREF data (see Section 2.4.7 for more details about these data), selected using `subset`. We call it as follows:

```
> library(boot)
> pf.pt.1 <- subset(pref.point, point==1)
> rownames(pf.pt.1) <- 1:nrow(pf.pt.1)
> pref.SRS.boot <- boot(pf.pt.1$vol.m3.ha,
+                       boot.mean,
+                       R = 1999)
```

Here we have chosen 1999 bootstrap replicates. Efron and Tibshirani (1993) suggest the use of up to 199 replicates for estimating standard errors and 1999 for estimating quantiles. The resulting object contains the output of the bootstrap call. We can examine it thus:

```
> pref.SRS.boot

ORDINARY NONPARAMETRIC BOOTSTRAP


Call:
boot(data = pf.pt.1$vol.m3.ha, statistic = boot.mean, R = 1999)


Bootstrap Statistics :
```

```
     original     bias     std. error
t1* 143.9130 0.5546451     23.98449
```

We can also plot the bootstrap object, as per Figure 3.1. In interpreting this graph, we are looking for approximate normality of the realizations in order for our bootstrap estimate to be most reliable. If we are satisfied, then the standard deviation (sd) of the estimated bootstrap means may be used as an estimate of the standard error of the parameter.

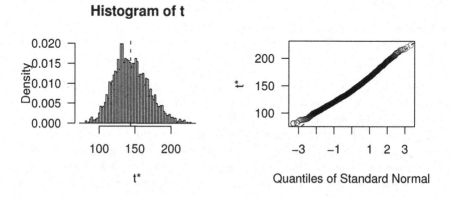

Fig. 3.1: Diagnostic graphical output for the bootstrap object, obtained by calling `plot(pref.SRS.boot)`.

```
> sd(pref.SRS.boot$t)
```

[1] 23.98449

The calculation of this estimate encapsulates the bootstrap innovation. We have computed an estimate directly from a simulated underlying distribution instead of an estimate that is based on an assumption about the underlying distribution. Much theory says that this is a defensible idea in many cases. It is certainly an improvement when the statistic of interest is a non-linear function of one or more parameters.

We can then either use the resulting estimate of the standard error directly to compute a confidence interval or we can use one of a number of other strategies that will be discussed shortly. The following approach assumes that the sampling distribution is normal, corrects for bias in the estimate of the mean, uses the standard deviation of the bootstrap estimates as the standard error, and uses the familiar 1.96 as a scale multiplier.

```
> boot.ci(pref.SRS.boot, type="norm")
```

```
BOOTSTRAP CONFIDENCE INTERVAL CALCULATIONS
Based on 1999 bootstrap replicates

CALL :
boot.ci(boot.out = pref.SRS.boot, type = "norm")

Intervals :
Level       Normal
95%    ( 96.3, 190.4 )
Calculations and Intervals on Original Scale
```

We can also deploy some graphical diagnostics to assess the effect of individual units upon our estimates. The jack.after.boot function presents for each observation the empirical distribution of the simulations that *omit* that observation (Figure 3.2). This diagnostic identifies rows 11 and 21 as having a strong influence on the results.

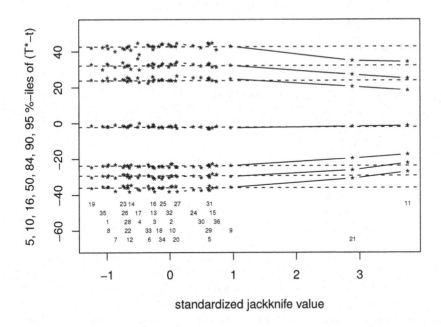

Fig. 3.2: Diagnostic graphical output of jackknifing the bootstrap object, obtained by calling jack.after.boot(pref.SRS.boot).

The identification of the rogue units is clarified by Figure 3.3, which provides a labeled normal quantile plot of the original sample, based on the qqnorm function. Figure 3.3 was constructed as follows:

```
> normalized <- qqnorm(pf.pt.1$vol.m3.ha, plot.it = FALSE)
> plot(normalized$x, normalized$y, type="n",
+      ylab="Sample Quantiles", xlab="Theoretical Quantiles",
+      main="Normal Q-Q Plot")
> qqline(pref.point$vol.m3.ha, col="darkgray")
> text(x = normalized$x, y = normalized$y, cex = 0.85)
```

The graphic suggests that rows 11 and 21 do not resemble the other rows. We can quickly confirm this by printing out the top five rows in terms of volume using the order function. Note the nesting of the subscript operations.

```
> pf.pt.1[order(pf.pt.1$vol.m3.ha, decreasing = TRUE),][1:5,]
```

	cluster	point	stratum	ba.m2.ha	vol.m3.ha	weight
11	15	1	3	64.27916	705.6219	1
21	31	1	6	91.82736	545.8158	1
9	13	1	3	45.91368	266.3664	1
5	7	1	2	55.09642	256.1642	1
31	47	1	8	55.09642	250.1831	1

The volumes in rows 11 and 21 are substantially higher than the others.

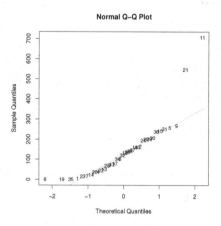

Fig. 3.3: Normal q-q plot of the PREF point-level volume data, replacing the points with the row numbers of the observations.

3.2.5.2 Innovations

We now consider different approaches to the construction of confidence intervals for a population parameter. As mentioned earlier, non-naive use of the bootstrap can lead to estimates that have better properties than those from naive use. For example, transforming the statistic of interest so that it is a pivot, also called a pivotal quantity, results in bootstrap simulations that have better properties (Davison and Hinkley, 1997).

The `boot.ci` function provides up to five different non-parametric confidence interval estimates for any given alpha value. The basic structure of the interval estimate is

$$\hat{\mu} + s_{\hat{\mu}} \times t_{\alpha/2}, \ \hat{\mu} + s_{\hat{\mu}} \times t_{1-\alpha/2} \tag{3.14}$$

The options for different intervals are as follows.

- norm: the normal confidence interval (3.14), with bootstrap-based bias correction for the estimate of the mean ($\hat{\mu}$), bootstrap estimate of the standard error $s_{\hat{\mu}}$, and z in place of t.
- stud: as per norm, but replacing t with the corresponding quantiles of the bootstrap sampling distribution of the studentized statistic. These are also called bootstrap-t intervals. The principle is to bootstrap a pivot; that is, a quantity whose distribution does not depend on the parameters. For example, the t statistic is a pivot for normal data because the distribution of t is unaffected by the mean and variance.
- basic: estimate the quantiles by subtracting the *opposite* observed quantiles from double the bootstrap estimate of the mean.[5]
- perc: directly use the quantiles of the estimated sampling distribution. This is the original percentile method, suggested by Efron.
- bca: adjusted percentile intervals, which are similar to the percentile intervals but change the quantiles to try to correct for bias.

The accuracy of the first three bootstrap estimators may improve for certain datasets if a variance-stabilizing transformation is employed. We will demonstrate the use of a variance-stabilizing transformation on page 90.

In order to studentize the simulations, it is necessary to obtain an estimate of the variance of the estimate within each realization. We could do this by adding another set of bootstrap operations within each bootstrap replicate or by using an approximation. Here we use an approximation. Since the statistic of interest is the sample mean, we can use the usual non-parametric estimator of the variance of the sample mean. This function must be computed and returned as part of the bootstrap function thus:

```
> bm.2 <- function(x, index)
+    c(mean(x[index]), var(x[index])/length(x))
```

---

[5] Manly (1997) calls this the second percentile method and attributes it to Peter Hall.

This function is then called as before, but now the bootstrap function can compute studentized values, which are approximate pivots.

```
> pref.SRS.b2 <- boot(pf.pt.1$vol.m3.ha, bm.2,
+                               R = 1999)
> pref.SRS.b2

ORDINARY NONPARAMETRIC BOOTSTRAP


Call:
boot(data = pf.pt.1$vol.m3.ha, statistic = bm.2, R = 1999)


Bootstrap Statistics :
       original        bias     std. error
t1*   143.9130    0.6177333      23.94001
t2*   589.5670  -10.3283157     265.16489
```

The bootstrap object now contains estimated means and variances. These can be graphed to assess stability. Here we impose a lowess curve on the scatterplot and hope that there is no pattern. Figure 3.4 shows a strong linear relationship between the bootstrap estimates of the mean and the variances.

```
> scatter.smooth(pref.SRS.b2$t[,1],
+                sqrt(pref.SRS.b2$t[,2]),
+                ylab = "Standard Error",
+                xlab = "Bootstrap Mean")
```

Since we know that the sample data are volumes and that they are skewed (and limited to positive numbers only), it is possible that a square root transformation will be useful. Figure 3.5 shows that using the transformation seems to be a reasonable strategy. Had we required a different transformation, then further searching would have been necessary.

```
> pref.SRS.b3 <- boot(I(sqrt(pf.pt.1$vol.m3.ha)),
+                      bm.2,
+                      R = 1999)
> scatter.smooth(pref.SRS.b3$t[,1],
+                sqrt(pref.SRS.b3$t[,2]),
+                ylab = "Standard Error",
+                xlab = "Bootstrap Mean")
```

A call to boot.ci, which includes appropriate forward and backward transformations, and the first derivative of the transformation, produces all five intervals.

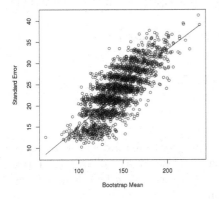

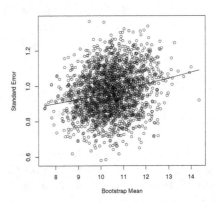

Fig. 3.4: Scatterplot and lowess smooth of estimated standard errors against estimated means from each of 1999 bootstrap replicates. The standard error and the mean are related.

Fig. 3.5: Scatterplot and lowess smooth of estimated standard errors against estimated means from each of 1999 bootstrap replicates taken from square root transformed data.

```
> h.dot <- function(x) 1/2 * x^(-1/2)
> h.inv <- function(x) x^2
> boot.ci(pref.SRS.b2,
+          h = sqrt,
+          hdot = h.dot,
+          hinv = h.inv)

BOOTSTRAP CONFIDENCE INTERVAL CALCULATIONS
Based on 1999 bootstrap replicates

CALL :
boot.ci(boot.out = pref.SRS.b2, h = sqrt, hdot = h.dot,
        hinv = h.inv)

Intervals :
Level      Normal           Basic         Studentized
95%    (101.3, 197.4)   ( 99.3, 196.2)   (103.5, 219.5)


Level      Percentile        BCa
95%    ( 99.7, 196.7)   (105.5, 205.8)
Calculations on Transformed Scale;   Intervals on Original Scale
```

The intervals are fairly similar, although the BCa and studentized intervals have longer right tails than the others.

It is presently unclear which of these intervals is the best to use. Davison and Hinkley (1997) recommend studentized intervals applied to data that have been transformed to stabilize variance.

We next report the results of a simulation study using a large-scale sample of forest inventory data as a population.

3.2.6 A Simulation Study

The best way to assess the different interval-generating strategies is to compare them in known situations on the basis of their realized coverage rates and relative lengths. Such a comparison requires complete knowledge of the target population and is rarely possible in a practical setting.

An alternative is to use ideal target populations; for example, the normal, the exponential, etc. The challenge is to find the subtle balance between reliably generating the natural patterns that provide realism and including sufficient randomness to provide a challenging test for the tool in question.

A slightly messier alternative is to obtain large databases that are within the scope of interest, treat the databases as the population of interest, and simulate the process of sampling. We report such an exercise in this section using the FIA data presented in Section 2.4.3.

Here we apply the ratio of means estimator, with the unweighted tree height as the auxiliary variable. This scenario is reasonable because tree heights have been successfully measured from aerial photographs and will become increasingly attractive as a source of auxiliary information with the introduction of remote measurement tools such as LIDAR. The relationship between height and basal area in our data is well suited to the use of the ratio of means estimator, being approximately linear, increasing in variance with height, and plausibly intersecting the origin (not shown here).

Each subpopulation was sampled randomly with replacement 2000 times, and the resultant sample data were used to estimate the mean stand basal area and a 95% confidence interval. We then counted whether the true value for the subpopulation was above, below, or within the confidence interval. The sample sizes used were 16, 32, 64, 96, and 128.

The average coverage rate is summarized by estimator and sample size in Figure 3.6.

```
> xyplot(contained ~ n | interval,
+        xlab = "Sample size (n)",
+        ylab = "Realized Coverage Rate",
+        panel = function(x, y) {
+          panel.xyplot(x, y, type="l")
+          panel.abline(h = 0.95, col = "darkgrey")
+        },
```

```
+        index.cond = list(c(3:5,10,1,9,14,2,6:8)),
+        skip = c(rep(FALSE, 11), TRUE, rep(FALSE, 3)),
+        data = ratio.test)
```

Note that an effective coverage of, say, 0.92, where 0.95 is expected, is analogous to behaving as though one collected a sample that was 1.12 times as many units as the actual sample size. Furthermore, an effective coverage of, say, 0.94, where 0.95 is expected, is analogous to behaving as though one collected a sample that was 1.04 times as many units as the actual sample size.

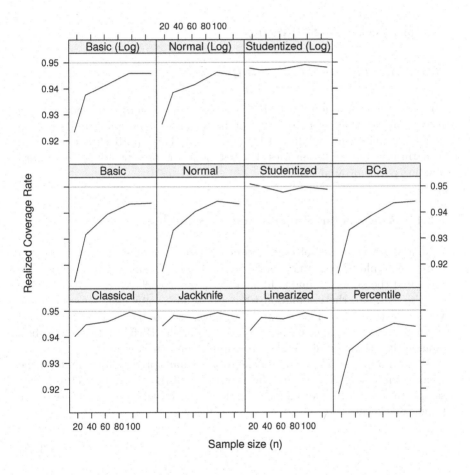

Fig. 3.6: Coverage rates against sample sizes for 14 different interval-estimating strategies.

Our results suggest that the coverage rate is reasonable for the classical, jackknife, and linearized estimators. The eight different bootstrap intervals performed poorly unless the sample was studentized or the sample size was sufficiently large, in which case it performed well. It is noteworthy that although bootstrapping the transformed data worked better than the untransformed data, studentizing was much more successful. Given that the linearization and jackknife tools are already available in R via the survey package, they seem suitable for our purposes.

Further examination of the results when categorized by nearest national forest did not reveal any important variation from these conclusions.

3.3 Single-Level Sampling

3.3.1 Simple Random Sampling

Simple random sampling involves the selection of a sample of known size n in such a way that every possible permutation of n sampling units has an equal and known probability of being selected. This approach is uncommon in natural resources inventories. It does, however, present a straightforward starting point for analysis.

3.3.1.1 Analysis for Simple Random Sampling

Assume that we have selected a simple random sample of n plots from a frame of N plot labels that completely represent a small forest stand. We measured the above-ground volume of the trees of each plot i, v_i m^2ha^{-1}. We wish to estimate the population total and a 95% confidence interval of the volume of the stand. For example, we have 180 observations of point-level above-ground volume in units of m^3ha^{-1} from the PREF dataset. We shall assume that these points represent a simple random sample.

We will obtain our estimate of the total by multiplying an estimate of the mean volume per hectare with the known size of the stand, in hectares. The sample mean of a simple random sample is a design-unbiased estimate of the population mean,

$$\bar{v} = \frac{1}{n}\sum_{i=1}^{n} v_i \tag{3.15}$$

The standard error is estimated using equation (3.3)

$$s_{\bar{v}} = \frac{s_v}{\sqrt{n}} \times \sqrt{1 - \frac{n}{N}} \tag{3.16}$$

where s_v is the standard deviation of the sample and the quantity $\sqrt{1 - \frac{n}{N}}$ is referred to as the finite population correction (FPC). The 95% confidence interval is then

$$95\% \text{ C.I.} = \bar{g} \pm s_{\bar{g}} \times t_{0.975, n-1} \qquad (3.17)$$

These computations can be done most efficiently in R via the survey package.

The data are in a data frame called **pref.point**, and the variable of interest is volume, reported in cubic meters per hectare, **vol.m3.ha**. We begin by declaring a survey object, which contains design information as well as the survey data. The required **id** argument identifies any hierarchical clustering, and the required **id** argument identifies the relative probability of selection of the sample units.

```
> library(survey)
> pref.SRS <- svydesign(id = ~1,
+                       data = pref.point,
+                       weight = pref.point$weight)
```

We note that the **svydesign** function requires a **weight** argument, which reports the sampling probability or the sampling weight of the observations. If a weight is not provided, then **svydesign** will issue a warning and assume equal weights. We now compute the estimate of the mean and its standard error as follows.

```
> svymean(~vol.m3.ha, pref.SRS)
```

```
            mean      SE
vol.m3.ha 141.62  8.8959
```

However, for simple random sampling, it is almost as easy to code the analysis manually.

```
> (SRS.v.hat <- mean(pref.point$vol.m3.ha))
```

```
[1] 141.6188
```

```
> SRS.n <- length(pref.point$vol.m3.ha)
> (SRS.v.se <- sd(pref.point$vol.m3.ha) / sqrt(SRS.n))
```

```
[1] 8.895931
```

```
> SRS.v.hat + SRS.v.se * qt(0.975, df = SRS.n - 1) * c(-1, 1)
```

```
[1] 124.0644 159.1732
```

The estimate of the total, and an appropriate confidence interval, can then be found by multiplying the estimate and confidence interval of the mean by the known forest area (2530 ha for PREF).

3.3.2 Systematic Sampling

Systematic sampling involves imposing a grid of sampling points with a one or more dimensions upon the population, preferably with random start and orientation. Generally speaking, systematic sampling provides good coverage of the population. An important exception is if one suspects that there is cyclic behavior in the population, in which case a systematic sample should not be used. An extreme example of such a spatial pattern is a plantation that has had different site preparations for different rows; for example, as might arise after slash mounding.

3.3.2.1 Analysis for Systematic Sampling

The Upper Flat Creek data are a systematic sample of 144 variable-radius plots. See Section 2.4.1 for a more detailed description of the sample design and the data collection methodology.

Estimating the population mean or total follows the same approach for systematic sampling as used for simple random sampling. We include the code here merely to set the scene for estimation of the standard error of the estimate.

```
> (SyRS.v.hat <- mean(ufc.SyRS.data$vol.m3.ha))
```

[1] 148.1544

Estimating the standard error is a little more complex for systematic random sampling because, in a design-based setting, the standard error is usually computed using the randomization distribution, and there is only one independent sample from that distribution. It is common that, for the purposes of variance estimation, samples from systematic random sampling are treated as though they were simple random samples. If the plot locations are known, or at least the order of the plot measurement is known, then better estimates of the variance are available.

Wolter (1985) provides an extensive discussion of eight distinct candidates, of which we select one that was also recommended by Schreuder et al. (1993) for use when nothing is known about the spatial structure of the population. The survey package does not currently support the use of spatial information for variance estimation in this way.

$$\hat{\sigma}_{\bar{y}} = \sqrt{\left(1 - \frac{n}{N}\right) \frac{1}{2n(n-1)} \sum_{i=2}^{n} (y_i - y_{i-1})^2} \qquad (3.18)$$

We omit the finite population correction because the sample units were points, not plots.

```
> SyRS.v.n <- length(ufc.SyRS.data$vol.m3.ha)
> vol.first.diffs <- ufc.SyRS.data$vol.m3.ha[1:(SyRS.v.n-1)] -
+   ufc.SyRS.data$vol.m3.ha[2:SyRS.v.n]
> SyRS.v.se <- sqrt(1 / (2 * SyRS.v.n * (SyRS.v.n - 1)) *
+                   sum(vol.first.diffs^2))
> SyRS.v.se
```

[1] 6.581681

The 95% confidence interval follows from the simple random sampling case. Had this sample been treated as a simple random sample, then the standard error would have been estimated as

```
> sd(ufc.SyRS.data$vol.m3.ha) / sqrt(SyRS.v.n)
```

[1] 8.106629

3.4 Hierarchical Sampling

Hierarchical sampling involves the selection of sets of sampling units, and possibly subsampling within those sets. Hierarchical sampling is less efficient than simple random sampling with the same number of sampling units. However, the inclusion in the budget of other elements of inventory, such as travel time, can make single-level sampling more expensive per unit of standard error. Furthermore, there are circumstances when it is impossible or prohibitively expensive to obtain a list of the sampling units in the population but easier or cheaper to obtain a list of clusters.

3.4.1 Cluster Sampling

Cluster sampling is a specific kind of hierarchical sampling in which sets of sampling units are selected and measured. As noted above, cluster sampling can be used to save on inventory costs, but it may also be used if a list of clusters is available but a sampling frame for the population is not. Cluster sampling can be considered as a special case of two-stage sampling (see Section 3.4.2), in which all of the secondary sampling units in each selected primary sampling unit are measured.

3.4.1.1 Analysis for Cluster Sampling

We have selected a simple random sample of n clusters from a frame of N clusters of plots, representing our population of interest. Each sampled cluster contained m_i plots, and we measured the volume for each plot, v_j. We wish to estimate the population total for the tree volume, τ_v, and a 95% confidence interval.

This is a cluster sample because the variable of interest is a plot-level characteristic, but the selection strategy uses only clusters of plots. Call the total of the measures in the i-th cluster t_i, so t_i is the sum of the plot volumes within cluster i. Denote the average number of plots per sampled cluster as $\bar{m}$.

For the PREF data, we have 36 clusters of five plots each. The data are contained in the `pref.point` object, the selection weight is in the `weight` variable, the cluster identifier is in the `cluster` variable, and the variable of interest is `vol.m3.ha`.

The estimate is

$$\bar{v}_c = \frac{\sum_{i=1}^{n} t_i}{\sum_{i=1}^{n} m_i} \tag{3.19}$$

In the survey package of R, the design is expressed by

```
> pref.CS <- svydesign(id = ~cluster,
+                       data = pref.point,
+                       weight = pref.point$weight)
```

and the estimate is

```
> svymean( ~ vol.m3.ha, pref.CS)

            mean      SE
vol.m3.ha 141.62 14.636
```

The variance of the estimator is the variance of the cluster means weighted by the number of units in each cluster,

$$s_{\bar{v}_c}^2 = \frac{N-n}{N} \frac{1}{n} \frac{1}{n-1} \sum_{i=1}^{n} \frac{m_i^2}{\bar{m}^2} \left(\frac{t_i}{m_i} - \bar{v}_c \right)^2 \tag{3.20}$$

We can easily compute both these classical estimates manually. The code that follows is intended to show the process rather than maximize the efficiency of the computation.

```
> v.clust <-
+   aggregate(x = list(vol.m3.ha = pref.point$vol.m3.ha),
+             by = list(cluster = pref.point$cluster),
+             FUN = sum)
> v.clust$count <- table(pref.point$cluster)
> n.clusters <- length(v.clust$count)
> (v.bar.c <- sum(v.clust$vol.m3.ha) / sum(v.clust$count))
```

```
[1] 141.6188
```

```
> (se.v.bar.c <-
+   sqrt(sum(v.clust$count^2 / mean(v.clust$count)^2 *
+           (v.clust$vol.m3.ha / v.clust$count -
+              v.bar.c)^2) / n.clusters / (n.clusters - 1)))
```

[1] 14.63561

When the clusters each have an identical number of sampling units, m, the variance estimate simplifies to

$$s_{\bar{v}_c}^2 = \frac{N-n}{N}\frac{1}{n}\frac{1}{n-1}\sum_{i=1}^{n}\left(\frac{t_i}{m}-\bar{v}_c\right)^2 \tag{3.21}$$

For the purposes of setting a confidence interval, we either need to pin down a reference distribution or apply the bootstrap. An exact reference distribution is difficult to pin down for cluster sampling. In practice, the following simplification is used:

$$95\% \text{ C.I.} = \bar{v}_c \pm s_{\bar{v}_c} \times 2 \tag{3.22}$$

If we wish to use the bootstrap for this problem, then we should probably bootstrap the clusters only (Davison and Hinkley, 1997, p. 101). Based on our experience in Section 3.2.6, we will studentize the bootstrap variable. The code for the mean and standard deviation can be used as above.

```
> boot.mean.cluster <- function(v.clust, index) {
+    v.clust <- v.clust[index,]
+    v.bar.c <- sum(v.clust$vol.m3.ha) / sum(v.clust$count)
+    c(v.bar.c,
+      sum(v.clust$count^2 / mean(v.clust$count)^2 *
+          (v.clust$vol.m3.ha / v.clust$count -
+             v.bar.c)^2) / n.clusters / (n.clusters - 1))
+ }
> pref.cluster.boot <- boot(v.clust,
+                           boot.mean.cluster,
+                           R = 1999)
> pref.cluster.boot

ORDINARY NONPARAMETRIC BOOTSTRAP


Call:
boot(data = v.clust, statistic = boot.mean.cluster, R = 1999)


Bootstrap Statistics :
     original    bias    std. error
```

```
t1* 141.6188 -0.1852036     14.45360
t2* 214.2011 -7.5990225     60.49281
```

```
> boot.ci(pref.cluster.boot, type = "stud")
```

```
BOOTSTRAP CONFIDENCE INTERVAL CALCULATIONS
Based on 1999 bootstrap replicates
```

```
CALL :
boot.ci(boot.out = pref.cluster.boot, type = "stud")
```

```
Intervals :
Level     Studentized
95%    (115.4, 177.2)
Calculations and Intervals on Original Scale
```

These are only slightly different from the classical approach as in equation (3.22):

```
> v.bar.c + c(-1, 1) * 2 * se.v.bar.c
```

```
[1] 112.3476 170.8900
```

3.4.2 Two-Stage Sampling

Two-stage sampling is similar to cluster sampling in that the units upon which the measurements are made are selected in groups, or clusters. However, the clusters are now referred to as primary sampling units (PSUs), and the sampling units within the clusters are called secondary sampling units (SSUs). Instead of measuring every sampling unit within the cluster, we now sample them. Here we will assume that sampling is with equal probability among the PSUs and SSUs and that the PSUs are all the same size.

3.4.2.1 Analysis for Two-Stage Sampling

We have collected a simple random sample of $n = 36$ primary sampling units (plots) from a frame of N labels of plots, representing our population. Each potential plot was a 90 m horizontal square projected onto the landscape, so approximately 3124 sampling units are possible in the 2530 ha.

There are M_i secondary sampling units, for example points, on each selected plot, of which we have measured m_i for volume v_{ij}. We wish to estimate the population mean and a 95% confidence interval. To keep the algebra simple, we will assume that the same number of points is measured in each plot;

i.e., that $m_i = m$. Also note that for variable-radius plot sampling, $M = \infty$, so we will use a very small second-stage FPC weight.

The estimate is

$$\bar{v}_{ts} = \frac{\sum_{i=1}^{n} \sum_{j=1}^{m} v_{ij}}{n \times m} \tag{3.23}$$

The analysis of such data in R is straightforward. We now declare two layers of hierarchy in our sampling units and proceed as before. We also add a variable to the dataset as the FPC for the PSU. The design is expressed as follows.

```
> pref.2SS <- svydesign(id = ~cluster + point,
+                       fpc = ~rep(36/3124, nrow(pref.point)) +
+                             rep(0.000001, nrow(pref.point)),
+                       weight = pref.point$weight,
+                       data = pref.point)
```

To obtain the estimates, we use

```
> svymean(~vol.m3.ha, pref.2SS, na.rm = TRUE)

            mean      SE
vol.m3.ha 141.62 14.569
```

This estimate is practically identical to that from cluster sampling.

The standard error of the mean is marginally more complex than for cluster sampling. The development is similar to that of analysis of variance in that the estimate of the standard error is the weighted sum of estimates of variances of the different levels of units. Each sampling book we referred to had a slightly different exposition of the variance; we found Cochran (1977) the clearest. Let $f_1 = n/N$ and $f_2 = m/M$. Then

$$s_{\bar{v}_{ts}}^2 = \frac{1 - f_1}{n} s_1^2 + \frac{f_1(1 - f_2)}{mn} s_2^2 \tag{3.24}$$

where s_1^2 is the variance of the means of the PSUs and s_2^2 is the variance of the SSUs within the PSUs.

$$s_1^2 = \frac{1}{n-1} \sum_{i=1}^{n} (\bar{v}_i - \bar{\bar{v}})^2 \tag{3.25}$$

$$s_2^2 = \frac{1}{n(m-1)} \sum_{i=1}^{n} \sum_{j=1}^{m_i} (v_{ij} - \bar{v}_i)^2 \tag{3.26}$$

Note that equation (3.25) is just the variance of the PSU means and equation (3.26) is the within-PSU variance. Computation proceeds as follows.

```
> v.2SS <-
+    aggregate(x = list(vol.m3.ha.bar = pref.point$vol.m3.ha),
```

```
+                    by = list(cluster = pref.point$cluster),
+                    FUN = mean)
> v.2SS$count <- table(pref.point$cluster)
> n.2SS <- nrow(v.2SS)
> N.2SS <- 3124
> m.2SS <- mean(v.2SS$count)
> M.2SS <- 10000
> f1.2SS <- n.2SS / N.2SS
> f2.2SS <- m.2SS / M.2SS
> pref.point.temp <- merge(pref.point, v.2SS)
> v1 <- var(v.2SS$vol.m3.ha.bar)
> v2 <- 1 / n.2SS / (m.2SS - 1) *
+    sum((pref.point.temp$vol.m3.ha -
+         pref.point.temp$vol.m3.ha.bar)^2)
> se.v.bar.2SS <- sqrt((1 - f1.2SS)/n.2SS*v1 +
+                       f1.2SS*(1-f2.2SS)/(n.2SS*m.2SS)*v2)
> se.v.bar.2SS
```

[1] 14.56936

If n/N is negligible, then it is reasonable to assume that

$$s^2_{\bar{v}_{ts}} \simeq \frac{s^2_1}{n} \qquad (3.27)$$

and in R

```
> sd(v.2SS$vol.m3.ha) / sqrt(n.2SS)
```

[1] 14.63561

Another development of the standard error is found in Avery and Burkhart (2003). We translate the development into terms more similar to those used by Cochran (1977) for ease of comparison. Again, let $f_1 = n/N$ and $f_2 = m/M$. Then, from Avery and Burkhart (2003, p. 61),

$$\begin{aligned}
s^2_{\bar{v}_{ts}} &= \frac{1}{nm}\left[s^2_B\left(1 - \frac{n}{N}\right) + \frac{ns^2_W}{N}\left(1 - \frac{m}{M}\right)\right] \\
&= \frac{1 - f_1}{mn}s^2_B + \frac{f_1(1 - f_2)}{mn}s^2_W
\end{aligned} \qquad (3.28)$$

where s^2_B is the variance of the means of the PSUs when estimated from m SSUs and s^2_W is the variance of the SSUs within the PSUs. s^2_W and s^2_B are defined as follows.

$$s^2_B = \frac{1}{n-1}\left(\frac{\sum_{i=1}^n\left(\sum_{j=1}^m v_{ij}\right)^2}{m} - \frac{\left(\sum_{i=1}^n\sum_{j=1}^m v_{ij}\right)^2}{mn}\right) \qquad (3.29)$$

$$s_W^2 = \frac{1}{n(m-1)} \left(\sum_{i=1}^{n} \sum_{j=1}^{m} v_{ij}^2 - \frac{\sum_{i=1}^{n} \left(\sum_{j=1}^{m} v_{ij} \right)^2}{m} \right) \tag{3.30}$$

Some brief algebra shows us that $s_1^2 = \frac{s_B^2}{m}$ and $s_2^2 = s_W^2$. The authors note that if n/N is negligible, then it is reasonable to assume that

$$s_{\bar{v}_{ts}}^2 \simeq \frac{s_B^2}{mn} \tag{3.31}$$

Note also that s_W^2 and s_B^2 can be conveniently obtained from the mean square column of an analysis of variance table.

```
> summary(aov.tab <- aov(vol.m3.ha ~ cluster, data=pref.point))

            Df  Sum Sq  Mean Sq  F value    Pr(>F)
cluster     35  1349467   38556   4.6254  3.163e-11 ***
Residuals  144  1200346    8336
---
Signif. codes:  0 '***' 0.001 '**' 0.01 '*' 0.05 '.' 0.1 ' ' 1

> unlist(aov.tab)[5:6]

$coefficients.cluster7
[1] 304.9582

$coefficients.cluster8
[1] 51.35599

> v1*5

[1] 38556.2

> v2

[1] 8335.738
```

These equations will yield results identical to those of Cochran (1977).

We note in passing that an incorrect estimator is presented in the third and fourth editions of Husch et al. (2003). The second edition presents correct formulas that result in estimates similar to those from Cochran (1977).

The final unique development of the standard error that we cover can be found in Schreuder et al. (1993). The authors suggest computing the within-PSU variance for each sampled PSU instead of pooling the variances, which is as the previous developments suggest. From Schreuder et al. (1993, equation (5.12)), the variance of the estimate of the *total* is

$$s_{\bar{y}_{ts}}^2 = N^2 M_a^2 (1-f) \frac{s_B^2}{n} + \frac{N}{n} \sum_{i=1}^{n} M_i^2 (1-f_i) \frac{s_{Wi}^2}{m_i} \qquad (3.32)$$

where $f = n/N$, $f_i = m_i/M_i$, $M_a = \sum_{i=1}^{N} M_i/N$, m_i is the number of SSUs selected in PSU i, and s_{Wi}^2 is the variance within PSU i. Also, the authors' label for our equation (3.25) is s_b^2 instead of s_1^2.

Substituting terms that we have used above, and assuming that the same number of SSUs is sampled in each PSU, the variance of the estimate of the *mean* is

$$s_{\bar{y}_{ts}}^2 = (1-f_1) \frac{s_1^2}{n} + \frac{f_1(1-f_2)}{n^2} \sum_{i=1}^{n} \frac{s_{Wi}^2}{m} \qquad (3.33)$$

which, if we assume $s_{Wi}^2 = s_W^2$, is identical to (3.24).

3.5 Using Auxiliary Information

The efficiency of estimation from sampling can be greatly enhanced by the use of auxiliary information. Auxiliary information is some kind of knowledge about the sampling units that is related to the variable of interest and that is known for *all* of the units in the population, or at least the population parameters are known. Auxiliary information can be used for the design of a sample (as per stratification, Section 3.5.1) or for the estimation (as per ratio and regression estimation, Sections 3.5.2 and 3.5.3, respectively).

3.5.1 Stratified Sampling

Stratification enables us to control variation. To stratify, we divide the population into discrete, non-overlapping subpopulations using an auxiliary variable. We then sample within each subpopulation without reference to the other subpopulations. That is, the sampling is carried out independently.

Stratification controls variation by allowing us to calculate separate statistics for things that we think will differ, a priori. We then compute population-level estimates by taking weighted averages of these stratum-level statistics. In this way, the variation *between* the subpopulations does not enter the uncertainty of the estimate.

We can stratify using a discrete or a continuous auxiliary variable. Once we have stratified, we do not use the auxiliary variable further unless we explicitly involve it by means of a more advanced estimation technique. In stratification, the auxiliary variable only affects the design. Once we have the design, it plays no further role.

3.5.1.1 Analysis for Stratified Sampling

Our population is still the PREF experimental forest (see Section 2.4.7). The forest was stratified into $L = 9$ strata, of relative size w_h, $h = 1 \ldots 9$, and we shall assume that simple random sampling was carried out independently within each stratum and that the strata are of equal size. We ignore the clustering of the actual sample for the moment. Each point was measured for volume, v_{ij} m^3/ha. We wish to estimate the population mean ($\bar{v}_{sts}$) and a 95% confidence interval for the population mean.

The estimate is

$$\bar{v}_{sts} = \sum_{h=1}^{L} w_h \bar{v}_h \qquad (3.34)$$

The survey package allows us to declare the strata as an argument. We have only to provide the name of a factor that describes stratum membership. The design is then

```
> pref.StRS <- svydesign(id = ~1,
+                        strata = ~stratum,
+                        data = pref.point,
+                        weight = pref.point$weight)
```

and the estimate is obtained by

```
> svymean(~vol.m3.ha, pref.StRS)

              mean      SE
vol.m3.ha  141.62  8.7787
```

This estimate is very similar to the previous estimates, and the standard error is lower, because the clustering has been ignored. We will analyze the design more appropriately shortly.

The variance of the mean is simply the weighted sum of the variances of the stratum-level means. This equation enables straightforward estimation from a stratified sample even if different sampling regimes have been used for each stratum.

$$s_{\bar{v}_{sts}} = \sqrt{\sum_{h=1}^{L} w_h^2 s_{\bar{v}_h}^2} \qquad (3.35)$$

We can still easily compute these estimates manually.

```
> stratum.weights <- rep(1/9, 9)
> v.str <- sum(stratum.weights *
+             tapply(pref.point$vol.m3.ha,
+                    pref.point$stratum,
+                    mean))
```

```
> se.v.str <- sqrt(sum(stratum.weights^2 *
+                     tapply(pref.point$vol.m3.ha,
+                          pref.point$stratum, var) /
+                     tapply(pref.point$vol.m3.ha,
+                          pref.point$stratum, length)))
```

Again, the exact reference distribution for constructing an appropriate confidence interval is not known; the common approximation is to subtract a degree of freedom from the t distribution for each stratum.

$$95\% \text{ C.I.} = \bar{v}_{sts} \pm s_{\bar{v}_{sts}} \times t_{n-L} \qquad (3.36)$$

3.5.1.2 Combinations of Designs

An advantage of the adoption of the survey routines now becomes obvious: the blending of simple design elements into more complicated designs is relatively straightforward. Thus, if we have stratified our population and performed cluster sampling within each stratum, then the appropriate code is as follows.

```
> pref.CStRS <- svydesign(id = ~cluster,
+                          strata = ~stratum,
+                          data = pref.point,
+                          weight = pref.point$weight)
> svymean(~vol.m3.ha, pref.CStRS)

            mean     SE
vol.m3.ha 141.62 15.527
```

Note that with the inclusion of clusters, the effective sample size within each stratum plummeted, and the standard error is actually larger than the estimate without stratification. This suggests that stratification was a waste of time for our purposes. However, it may well have proven to be more useful if we had been interested in estimates of volume by species.

3.5.2 Ratio Estimation

When our auxiliary variable is continuous, then it may be more efficient to involve our knowledge of the variable directly in the estimation process. We consider two principal strategies for this approach: ratio estimation and regression estimation.

Ratio estimation itself has two variants, labeled the ratio of means and the mean of ratios. The latter is mostly applied when sampling is performed with probability proportional to the auxiliary variable. The variable of interest from the point of view of sampling is then the ratio of the variable of interest and the auxiliary variable, and computation proceeds according to variable probability sampling.

Here we focus on the ratio of means case. Ratio of means estimation is appropriate if we are reasonably certain that the relationship between the variable of interest and the auxiliary variable is linear and passes through the origin and that the variance of the variable of interest is roughly proportional to the value of the auxiliary variable.

This case provides the simplest interesting challenge to the sampler, as the population parameter of interest is a non-linear function of two statistics. As we have discussed earlier in the context of interval estimation, estimating the variance is complicated.

3.5.2.1 Analysis for Ratio Estimation

Continuing the analysis of the PREF inventory data, we now introduce the remotely sensed auxiliary variable, which is the atmospherically corrected Normalized Difference Vegetation Index (NDVI), adjusted using the mid-infrared band. This variable is available for every 30 m by 30 m pixel of the forest and can be expected to be fairly well correlated with the per-unit volume. As previously, the use of the survey package simplifies the analysis considerably.

As noted earlier in the chapter, the following two equations provide the usual estimate of the population mean of the variable of interest and an estimate of its standard error. Initially, we will ignore some of the important elements of the PREF design: the clustering and the stratification.

$$\hat{\mu}_y = \mu_x \times \frac{\bar{y}}{\bar{x}} \tag{3.37}$$

$$s_{\hat{\mu}_y}^2 = \frac{\mu_x^2}{\bar{x}^2} \times \frac{s_y^2 + \hat{R}^2 s_x^2 - 2\hat{R}s_{xy}}{n} \tag{3.38}$$

We register the design as follows

```
> pref.SRSc <- svydesign(id = ~1,
+                        data = pref.point.cov,
+                        weight = pref.point$weight)
```

Then the analysis is performed by

```
> volume.over.acndviC <- svyratio(numerator = ~vol.m3.ha,
+                                  denominator = ~acndviC,
+                                  design = pref.SRSc)
> predict(volume.over.acndviC, mean(pref.pixel$ndvic))

$total
          acndviC
```

```
vol.m3.ha 139.5000

$se
          acndviC
vol.m3.ha 8.317442
```

We can easily reproduce these straightforward results using our own code to
check our understanding.

```
> ratio.hat <- sum(pref.point.cov$vol.m3.ha) /
+    sum(pref.point.cov$acndviC)
> RatE.v.hat <- mean(pref.pixel$ndvic) * ratio.hat
> RatE.v.hat

[1] 139.5000

> RatE.v.hat.se <-
+    sqrt((mean(pref.pixel$ndvic))^2 /
+         (mean(pref.point.cov$acndviC))^2 *
+         (var(pref.point.cov$vol.m3.ha) +
+          ratio.hat^2 * var(pref.point.cov$acndviC) -
+          2 * ratio.hat * cov(pref.point.cov$acndviC,
+                              pref.point.cov$vol.m3.ha)) /
+         nrow(pref.point.cov))
> RatE.v.hat.se

[1] 8.317442
```

The results of the simulation study reported in Section 3.2.6 suggest that
the use of the jackknife or the linearized estimates of the standard error,
along with the t quantile that is appropriate for the desired coverage, may
provide a reasonable confidence interval.

3.5.2.2 Combinations of Designs

As before, the use of the survey package permits the straightforward extension of these simple design elements into a more complex whole. Here we
include the design elements of stratification and clustering as well as the
ratio estimation in our analysis. The design is

```
> pref.CStRSc <- svydesign(id = ~cluster+point,
+                          strata = ~stratum,
+                          data = pref.point.cov,
+                          weight = pref.point$weight)
```

and the analysis requires the following code:

```
> v.over.a.CStRSc <- svyratio(numerator = ~vol.m3.ha,
+                              denominator = ~acndviC,
+                              design=pref.CStRSc)
> predict(v.over.a.CStRSc, mean(pref.pixel$ndvic))

$total
            acndviC
vol.m3.ha 139.5000

$se
            acndviC
vol.m3.ha 13.77745
```

This analysis is closest to being the most appropriate analysis for the original design.

3.5.3 Regression Estimation

In the more general case when we are uncertain as to the nature of the relationship between the variable of interest and the auxiliary variable, specifically the form of the conditional mean and variance, we can apply regression estimation.

3.5.3.1 Analysis for Regression Estimation

Continuing the analysis of the PREF inventory data, we again use the remotely sensed auxiliary variable, which is the atmospherically corrected Normalized Difference Vegetation Index, adjusted using the mid-infrared band. This variable is available for every 30 m by 30 m pixel of the forest and can be expected to be fairly well correlated with the per-unit volume. Again, the use of the survey package simplifies the analysis considerably.

The following four equations provide the usual estimate of the population mean of the variable of interest and an estimate of its standard error. Initially we again ignore the important elements of the PREF design: the clustering and the stratification.

The regression-based estimate of the mean of the variable of interest, y, is

$$\bar{y}_{lr} = \bar{y} + \hat{\beta}_1 \times (\bar{X} - \bar{x}) \tag{3.39}$$

where $\bar{x}$ is the mean of the auxiliary variable for the sample $\bar{X}$ is the (known) mean of the auxiliary variable for the population, and

$$\hat{\beta}_1 = \frac{s_{xy}}{s_x^2} \tag{3.40}$$

The standard error is estimated as follows:

$$s_{y|x} = \sqrt{\frac{n-1}{n-2}\left(s_y^2 - \frac{s_{xy}^2}{s_x^2}\right)} \tag{3.41}$$

$$s_{\bar{y}_{lr}} = s_{y|x} \times \sqrt{\left(\frac{1}{n} + \frac{(n-1)(\bar{X}-\bar{x})^2}{s_x^2}\right) \times \left(1 - \frac{n}{N}\right)} \tag{3.42}$$

These estimates are computed in R using the following code:

```
> volume.against.acndviC <- svyglm(vol.m3.ha ~ acndviC,
+                                     design = pref.SRSc)
```

We now rely on another package, gmodels (Warnes, 2010), to provide us with appropriate predictions and standard errors. We will use the estimable function for this purpose. This very useful function will compute estimates, standard errors, confidence intervals, and hypothesis tests for arbitrary linear combinations of parameters from a wide variety of model types, including mixed-effects models. We load the package using the library function.

```
> library(gmodels)
```

We then invoke the estimable function as follows.

```
> estimable(volume.against.acndviC,
+           conf.int = 0.95,
+           cm = rbind(total=c(1, mean(pref.pixel$ndvic))))

      Estimate Std. Error   t value  DF Pr(>|t|) Lower.CI
total 139.3812    8.34162  16.70913 178        0 122.9200
      Upper.CI
total 155.8424
```

3.5.3.2 Combinations of Designs

Again we can draw upon the survey package to provide us with a general yet straightforward solution that is appropriate for the full PREF design.

```
> pref.CStRSc <- svydesign(id = ~cluster+point,
+                          strata = ~stratum,
+                          weight = pref.point$weight,
+                          data = pref.point.cov)
> v.against.a.CStRSc <- svyglm(vol.m3.ha ~ acndviC,
+                              design = pref.CStRSc)
> estimable(v.against.a.CStRSc,
+           conf.int = 0.95,
+           cm = rbind(total = c(1, mean(pref.pixel$ndvic))))
```

```
       Estimate Std. Error  t value DF     Pr(>|t|) Lower.CI
total 139.3812    13.66037 10.20332 26 1.391884e-10 111.3019
       Upper.CI
total 167.4605
```

Note that the acknowledgment of the clustering of sampling units has inflated the estimate of the standard error considerably.

3.5.4 3P Sampling

3P sampling, also called *Poisson sampling* (see, e.g., Schreuder et al., 1993), is a system that involves guessing the variable of interest before measuring it, and using the guesses as auxiliary information.

3.5.4.1 Analysis for 3P Sampling

We have a population of N trees, and we wish to estimate the total volume. We have guessed the population size N and generated N random numbers, q_i, uniformly distributed from 0 to q_{max}. q_{max} corresponds to the estimated total divided by the desired sample size.

We visited each of the trees and guessed the volume in cubic meters, x_i. If $q_i < x_i$, we measured the tree accurately for volume y_i. The guesses sum to X_t for the whole population. We wish to estimate the population total and a 95% confidence interval.

For the purposes of demonstration, we will use the PREF tree database as our population. The variable of interest is now volume in cubic meters, vol.m3. We set up the sample as follows. We believe that the tree count in the population is 1000, and we wish to accurately measure, say, 20 trees. Further, we shall guess that the volume of all the trees in the forest is, say, 1000 m^3. Then we generate 1000 random numbers between 0 and $1000/20 = 50$.

```
> q.max <- 1000/20
> PPP.rand <- q.max * runif(1200)
```

The guesses can be simulated by either adding a random variable to or multiplying a random variable by the known values. All kinds of scenarios can be compared. The authors are notoriously poor at guessing tree volumes, so for the purpose of generating realistic data to demonstrate the technique,

```
> PPP.guesses <- pref.tree$vol.m3 *
+   rnorm(n = nrow(pref.tree), mean = 1.05, sd = 0.15) +
+   rnorm(n = nrow(pref.tree), mean = -0.05, sd = 0.05)
```

We then choose the sample using one of numerous possible approaches.

```
> PPP.sample <-
+    PPP.rand[seq(1,length(PPP.guesses))] < PPP.guesses
> PPP.trees <-
+    data.frame(vol.m3 = pref.tree$vol.m3[PPP.sample],
+                guess.m3 = PPP.guesses[PPP.sample])
```

The estimate is then

$$\hat{Y}_{3P_*} = X_t \times \frac{1}{n} \times \left(\sum_{i=1}^{n} r_i \right) \tag{3.43}$$

$$r_i = \frac{y_i}{x_i} \tag{3.44}$$

and in R

```
> (PPP.hat.1 <- sum(PPP.guesses) / sum(PPP.sample) *
+    sum(PPP.trees$vol.m3 / PPP.trees$guess.m3))
```

[1] 1217.149

An alternative estimator with slightly better properties, in that it corrects somewhat for the randomness of the sample size (Schreuder et al., 1993), is

$$\hat{Y}_{3P} = \frac{1}{n} \times \frac{X_t}{q_{max}} \times \hat{Y}_{3P_*} \tag{3.45}$$

In R,

```
> (PPP.hat.2 <- PPP.hat.1 *
+    sum(PPP.guesses) / q.max / sum(PPP.sample))
```

[1] 905.2569

The actual value is known to be

```
> sum(pref.tree$vol.m3)
```

[1] 1185.261

The standard error of the estimate also has numerous incarnations. A simple version is

$$s_{\hat{Y}_{3P}} = X_t \times s_{\bar{r}} \tag{3.46}$$

```
> se.PPP.hat <- sd(PPP.trees$vol.m3 / PPP.trees$guess.m3) /
+    sqrt(sum(PPP.sample)) * sum(PPP.guesses)
> se.PPP.hat
```

[1] 44.86651

As with other complicated sampling routines, the reference value for the sampling distribution of the statistic is not known. The usual approximation is used,

$$95\% \text{ C.I.} = \hat{Y}_{3P} \pm 2 \times s_{\hat{Y}_{3P}} \tag{3.47}$$

3.5.5 VBAR

VBAR stands for variable-basal area ratio. It is an efficient way to gather dimensional information on trees in a stand. It is most often applied using variable-radius plots, but there is no reason that it could not be applied with fixed-area plots.

The idea behind VBAR is to use many cheap basal area plots to cover the spatial variation and a smaller number of more expensive plots to establish a relationship between the basal area and the variable of interest. The two best references for this technique are Bell and Dillworth (1997) and Iles (2003).

VBAR was originally coined as volume-basal area ratio, but it has since (rightly) been pointed out that the technique can be used for measuring most anything, as long as it is likely to be related to basal area at the point level, by some kind of ratio. It is formally known as two-phase sampling in statistical literature.

3.5.5.1 Analysis for VBAR

We have installed n_1 variable-radius plots. In each one, we only counted the trees. Among these plots, we took a subsample of measure plots, upon which we ended up measuring n_2 trees for basal area g_i and volume v_i. We wish to estimate the population total and a 95% confidence interval.

We can easily alter the PREF database to demonstrate the use of VBAR. We have already computed the basal area and volume from each plot, so we just need to choose a sample of the plot numbers. We shall take a sample of 30 volume plots.

```
> vbar.sample <- sample(1:dim(pref.point)[1], size=30)
> vbar.points <- pref.point[vbar.sample,]
> vbar.points <- vbar.points[vbar.points$ba.m2.ha > 0,]
> vbar.points$ratio <- vbar.points$vol.m3.ha /
+    vbar.points$ba.m2.ha
```

Empty plots should be eliminated from the sample for the purposes of computing the ratio and its standard error. Note that this is a rare occasion when ignoring measured plots is reasonable. Our rationale is that 0/0 is undefined. Also note that this strategy effectively makes the sample size for the ratio a random variable, which, if correctly accounted for, will complicate the analysis. We ignore that element of the analysis here. Probably the simplest approach to accommodating this element of the design would require use of the bootstrap.

The estimate is

$$\bar{v}_{VBAR} = \bar{g} \times \overline{v/g} \qquad (3.48)$$

```
> (v.hat.vbar <-
+   mean(pref.point$ba.m2.ha) * mean(vbar.points$ratio))
```

```
[1] 138.6455
```

The units are, as before, m^3/ha.

The estimation of standard error follows principles laid out in Goodman (1960),

$$s_{\bar{v}\%} = \sqrt{s_{\bar{g}\%}^2 + s_{\overline{v/g}\%}^2 - s_{\bar{g}\%}^2 \times s_{\overline{v/g}\%}^2} \qquad (3.49)$$

Here $s_{\bar{v}\%}$ is the ratio of the standard error of the volume and the mean volume, $s_{\bar{g}\%}^2$ is the square of the ratio of the standard error of the basal area and the mean basal area, and $s_{\overline{v/g}\%}^2$ is the square of the ratio of the standard error of the ratios and the mean ratio,

$$s_{\bar{v}_{VBAR}} = \bar{v}_{VBAR} \times s_{\bar{v}\%} \qquad (3.50)$$

```
> s.bar.g.perc.2 <- (sd(pref.point$ba.m2.ha) /
+                    sqrt(length(pref.point$ba.m2.ha)) /
+                    mean(pref.point$ba.m2.ha))^2
> s.bar.r.perc.2 <- (sd(vbar.points$ratio) /
+                    sqrt(length(vbar.points$ratio)) /
+                    mean(vbar.points$ratio))^2
> s.bar.v.perc <- sqrt(s.bar.g.perc.2 + s.bar.r.perc.2 -
+                    s.bar.g.perc.2 * s.bar.r.perc.2)
> s.bar.v <- s.bar.v.perc * v.hat.vbar
> s.bar.v
```

```
[1] 9.71663
```

The confidence interval again uses 2 as the approximate quantile of the sampling distribution.

3.6 Summary

This chapter gives a brief presentation of the tools available for sample processing and confidence interval estimation. We chose to focus on the operations required for the various sampling designs normally found in forest inventories, but the reader should be aware that these methods can apply to any situation where these sampling designs are applicable.

Chapter 4
Imputation and Interpolation

4.1 Introduction

In Chapter 3, we presented methods to process samples, estimate parameters, and construct confidence intervals for design-based inference. In this chapter, we present model-based imputation (to fill in missing values) and interpolation (for predicting values at unsampled locations) methods to generate complete datasets so that 1) we have no missing values in our analysis dataset or so that 2) we have complete coverage using predicted values at unsampled locations for some variable of interest.

In Section 4.2, we impute missing values for the `stands` data frame object. We use definitions consistent with those presented by Little and Rubin (2002) and define *imputation* as any process that replaces missing values (`NA`) with other predicted or observed values. We first examine the `stands` data to determine if any data are missing. Then, we examine missingness patterns to determine if the missingness has a simple spatial pattern. We then present a few methods to impute missing values in Section 4.2.2. In Sections 4.2.3 and 4.2.4, we impute missing values using the nearest-neighbor (NN) and the expectation-maximization (EM) algorithms. Finally, in Section 4.2.5, we briefly examine and compare the two results against the original data for a single site productivity variable, specifically site index (Bruce, 1981).

In Section 4.3, we then interpolate site index using the processed `plots` and `tree` data frame objects from Section 2.4.6. In Section 4.3.1, we present a few interpolation methods. In Section 4.3.2, we answer a few critical questions, select one interpolation method (specifically, kriging) and briefly present the theory. In Section 4.3.3, we estimate the variogram used in Section 4.3.4, where we finally predict site index over a finite region.

4.2 Imputation

Imputation is a process of filling in missing values. Imputation is often performed when complete data are required but only incomplete data are available. The specific protocols adopted usually depend on the subfield (e.g., inventory, growth and yield, simulation, and optimization). Imputation has received considerable attention in the forestry literature (e.g., Moeur and Stage, 1995; Holmström, 2002; Ohmann and Gregory, 2002; Holmström et al., 2003); however, we refer the reader to Little and Rubin (2002) for a more complete and general treatment of imputation methods. Here we use definitions that are consistent with those presented by Little and Rubin (2002) and simply define *imputation* as any process that replaces missing values (in R, NA) with other predicted or observed values.

Here we first examine the stands data using a variety of techniques to determine if data are missing and perform a few simple tests to determine if there is a spatial pattern to the missing data. In Section 4.2.2, we present methods to impute missing data and provide two examples, k-NN and EM, in Sections 4.2.3 and 4.2.4, respectively. Finally, in Section 4.2.5, we briefly compare the results of the two methods.

4.2.1 Examining Missingness Patterns

To quickly determine if missing values exist in the stands data, use the show.cols.with.na function, telling R to treat the stands object as a data frame.

```
> source("../../scripts/functions.R")
> stands <- readShapePoly("../../data/stands.shp",
+                         verbose = FALSE)

> show.cols.with.na(as.data.frame(stands))
```

STANDID	TAGE	BHAGE	DF_SITE	TOTHT	CUBVOL_AC
6	28	181	36	97	103
TPA	QMD	BA			
97	97	101			

The output reveals that many stand polygons contain some missing values. Recall from Chapter 2 that NA values represent an unobserved or missing value, not an observation made where the value of the observation was zero.

Graphical examination of missingness can also be quickly performed by generating a map of polygons shaded by the number of missing attributes. To generate that map, first compute the number of missing attributes for each stand polygon and append a MISSING attribute to the stands object,

```
> stands.frame <- as.data.frame(stands)
> stands.frame$MISSING <- rowSums(is.na(stands.frame))
```

The stands object now contains stands$MISSING, which reports the number of missing attributes for each stand polygon. Next, create a color ramp from the number of unique values of the MISSING attribute:

```
> brks <- sort(unique(stands.frame$MISSING))
> colors <- gray(length(brks):1 / (length(brks)))
```

Finally, plot the stands, add a title, and include a legend to produce Figure 4.1.

```
> plot(stands,
+        col = colors[findInterval(stands.frame$MISSING,
+                                   brks, all.inside = TRUE)],
+        forcefill = FALSE,
+        axes = TRUE)
> title(main="Attribute Missingness by Polygon")
> legend(1280000, 365000, brks, fill = colors, cex = 0.7,
+        ncol = 3, title = "# of Missing Attributes")
```

Figure 4.1 reveals a potential pattern in which missingness might be more prominent in the northern portion of the landscape than in the southern part. If the headquarters is located at or near the southern portion of the forest, for example, a lengthy drive might explain the pattern where the stands in the north portion of the forest might be sampled less frequently or have fewer measurements taken on fewer variables. Regardless of the mechanism that created the differences in missingness, the summaries presented above and Figure 4.1 show that 1) data are missing in the stands object; 2) some stand polygons have more missing attributes than others; and 3) there may be a spatial pattern; with more missing values in the north than in the south.

While this quick analysis cannot tell us for certain that there is a pattern to the missingness, missing data problems can be more formally expressed so that more sophisticated techniques can be applied to construct a complete dataset.

Let $\mathbf{Y}$ denote the $n \times m$ random matrix that represents the complete data matrix of n rows (*observations*) and m columns (*variables*) and follows a multivariate density function $f(y; \theta)$, where y is a particular realization of the random variable $\mathbf{Y}$ (*complete data*) and θ is a parameter vector that defines the distribution of y. Also, $\mathbf{Y}$ can be partitioned into a matrix of the non-missing observations Y_{ij}^{o} and missing observations Y_{ij}^{m}.

To describe missingness, let $\mathbf{M} \in (0, 1)$ be a random indicator matrix that represents the missingness of the data for the i-th observation on the j-th variable. When y_{ij} is missing, $\mathbf{M}_{ij} = 1$; otherwise, $\mathbf{M}_{ij} = 0$. The realization of $\mathbf{M}$, m, shows the pattern of missing values of y, a realization of Y that follows $f(y; \phi)$, where $f(y; \phi)$ defines the pattern of missing data m. The primary

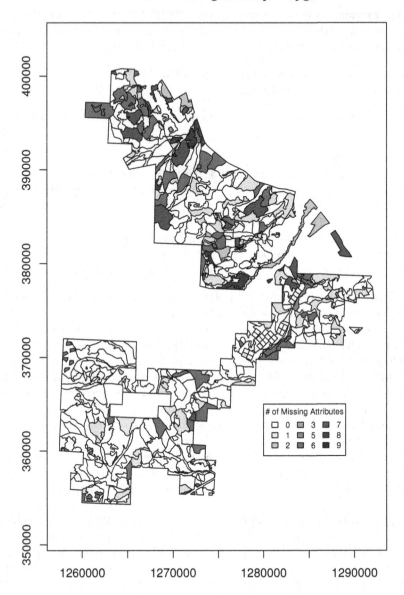

Fig. 4.1: This figure presents the stand polygons, where the darkness is a function of the number of missing attributes.

objective in most imputation problems is to describe $f(y; \phi)$ sufficiently so
that all pertinent missing values can be imputed to construct a complete
dataset.

Often, the indicator matrix **M** can be described by answering a few important contextual questions, such as:

1. What were/are the sampling design, plot design, and sampling frequency?
2. Is there any information about censoring observations in the field manuals
   or protocol documents?

Once these types of questions have been addressed, it is then possible
to more rigorously examine the data for patterns of missingness using the
techniques described by Little and Rubin (2002). Formally, the challenge is
to determine the values for θ that best define f, so you can say with some
certainty whether the data are missing with any detectable pattern, or if the
pattern is indeed random. Here we do not make an attempt to determine the
distinct missingness mechanism, and instead focus on the examination of m
itself.

First, we examine m by simply printing out the unique binary combinations
of m in the data. Since there are 13 columns in the **stands** dataset, there are
8192 possible (2^{13}) missingness patterns.

To create a simple table to examine row-wise summaries for m, use the
na.pattern function provided in the Hmisc package,

```
> nap <- na.pattern(stands.frame)
```

The **na.pattern** function provides a summary of the row-wise vectors for
m. Since printing the resulting object by itself is difficult to read, coercing
the resulting object into a matrix and then printing the results to obtain a
more presentable pattern yields

```
> as.matrix(nap)

                  [,1]
00000000000000    324
00000000000010      2
00000000111110      1
00000001111110      2
00000010000000     68
00000010000010      1
00000010010000      5
00000010111110     61
00000011000000      6
00000011000010      1
00000011111110     11
00000110000000      1
00000110010000      1
```

```
00000110111110    10
00000111000000    4
00000111111110    6
00010111111110    6
```

The call to convert the object to a matrix prints a frequency table of the combinations of missingness patterns for the `stands` object. In this case, there are 324 rows that have no missing attributes. There are 26 rows that have a 0000010111110 pattern, which is to say that there are 61 rows with the 6th and the 8th through the 13th variables missing. Again, since there are 13 variables for each row in the `stands` data frame, the total possible number of missing combinations is 2^{13}, or 8192, combinations, of which we only have 17, a considerable reduction in unique patterns. While this is interesting, it may not be very useful for generating a complete dataset. For additional diagnosis, however, use the `naclus` and `naplot` functions to examine m for possible relationships. Here, we only plot the patterns for visual inspection in Figure 4.2.

```
> par(mfcol=c(2,2), cex=0.7)
> nac <- naclus(stands.frame)
> naplot(nac)
```

Examination of Figure 4.2 reveals different missingness ratios. Specifically, the fractions of missingness values for DF_SITE ($\sum_{i=1}^{N} M_{i,j=7}$) and TAGE ($\sum_{i=1}^{N} M_{i,j=5}$) are similar. However, the variable with the highest fraction of NA values is the breast-height age (BHAGE), which is often associated with total age (TAGE) observations in this dataset. For example, it may be operational procedure for this dataset to sample a stand polygon only when enough time has passed so that all the stems have grown taller than 1.37 m (4.5 feet). Since total age can be updated once seedlings are planted, this might account for the difference in missingness. While BHAGE yields a large number of missing values when compared with the other variables, this may be one of those cases where having NA as a legitimate value can cause problems in statistical analysis, and we recommend caution.

Finally, we present another simple procedure to detect possible missingness patterns by predicting the probability of missingness using some metric that is easy to compute. In this case, we attempt to predict the probability of missingness using the Euclidean distance between the stand centroid and the forest headquarters (`stand.dist`). Then, using a Bernoulli response variable (0/1) to represent the presence or absence of a value (`site.missing <- is.na(DF_SITE)`), we can then use the `glm` function to fit a model of missingness.

First, compute the distance for each stand centroid to the headquarters, which is located, in rectangular coordinates, at 1288538.5625 east, 373896.78125 north. Using the `rdist` function, from the fields package (Furrer et al., 2009), we compute the distances from each of the stand centroids to the headquarters,

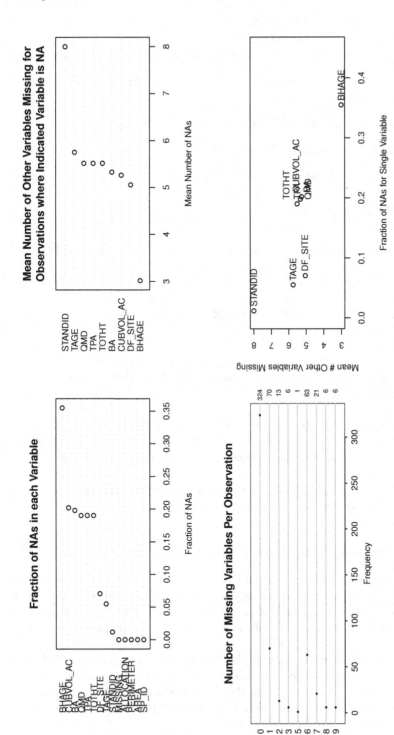

Fig. 4.2: Diagnostic plots examining the missing values for the variables in the **stands** data frame object.

```
> hq <- c(1288538.5625,373896.78125)
> centers <- coordinates(stands)
> stand.dist <- rdist(t(c(1288538.5625,373896.78125)), centers)
```

Then, we define a vector where the entries are NA if the site index is missing using the is.na function:

```
> site.na <- as.numeric(is.na(stands.frame$DF_SITE))
```

Finally, fit a binomial model using the glm function:

```
> gfit.site <- glm(site.na ~ c(stand.dist),
+                     family = "binomial")
> summary(gfit.site)

Call:
glm(formula = site.na ~ c(stand.dist), family = "binomial")

Deviance Residuals:
    Min       1Q   Median       3Q      Max
-0.4511  -0.4126  -0.3764  -0.3389   2.4567

Coefficients:
                 Estimate Std. Error z value Pr(>|z|)
(Intercept)    -3.015e+00  4.315e-01  -6.986 2.83e-12 ***
c(stand.dist)   2.145e-05  1.865e-05   1.150     0.25
---
Signif. codes:  0 '***' 0.001 '**' 0.01 '*' 0.05 '.' 0.1 ' ' 1

(Dispersion parameter for binomial family taken to be 1)

    Null deviance: 260.26  on 509   degrees of freedom
Residual deviance: 258.91  on 508   degrees of freedom
AIC: 262.91

Number of Fisher Scoring iterations: 5
```

The resulting p-value of 0.25 for the predictor variable suggests that there is little correlation between the distance of the stand centroid to the headquarters and the probability of the DF_SITE observation being missing. It seems that distance from the office is not a good predictor of missingness here.

We could continue to examine the dataset with more sophisticated tools and plots, but, in practice, the determination of the mechanism, and the method used to correct for it, is at the discretion of the analyst and judgment will dictate the final determination once our attempt to document the search for a missingness pattern is complete.

4.2.2 Methods for Imputing Missing Data

Now that we have determined that some data are missing, we present a few methods for imputation. Little and Rubin (2002) grouped the methods for imputing unobserved variables into three categories: 1) weighting procedures; 2) imputation-based procedures; and 3) model-based procedures. We briefly present them in the next few sections.

4.2.2.1 Weighting Procedures

When the sample selection is based on unequal probabilities of inclusion (sample weights), as described by Horvitz and Thompson (1952), then weighting procedures can adjust the weights used to compute various estimators (Little and Rubin, 2002, Chapter 3).

For relatively small amounts of missing data, weighting procedures that are based on completely recorded units (that is, with no missing values) can yield sufficient results. They can yield biased results and may not be very efficient when making inferences to subpopulations such as stands or strata (Little and Rubin, 2002, Chapter 3). In the current example, we believe we have sufficiently large numbers of missing observations in the stands object to give us biased results if we were to apply these methods. We do not present them here.

4.2.2.2 Imputation-Based Procedures

Imputation-based methods include those methods most commonly used in forestry, such as k-nearest neighbor (k-NN) and most similar neighbor (MSN) (Moeur and Stage, 1995), as well as less sophisticated methods such as *mean* and *regression* imputation methods . The *mean* imputation method simply replaces the missing attributes with the mean of the sampled attributes. The *regression* imputation method estimates unsampled attributes from sampled attributes using regression techniques (Little and Rubin, 2002, Chapter 1). These *hot-deck* methods of imputation are covered more completely elsewhere (e.g., by Venables and Ripley, 2002) and are not presented further here.

Alternatively, *cold-deck* methods assign values based on a constant value from an external source rather than from samples of the observed data. For example, in landscape-planning projects, regeneration stands might not have an adequate sample to make inference for the entire population over the span of the planning horizon. A decision is made to assign all unsampled stands that are under some arbitrary age a set of values that is "representative" of young stands across the landscape. This method ignores the consequences of imputation, such as bias, and is discouraged when unbiased results are desired (Little and Rubin, 2002, Chapter 4).

Finally, the nearest-neighbor methods are those for which unsampled measures are assigned values from nearby sampled attributes using a variety of statistical distances (Bishop, 2006, see, e.g., Chapter 2). We present our first imputation method, the k-NN method, in Section 4.2.3.

4.2.2.3 Model-Based Procedures

Model-based procedures are methods that make inferences upon the distribution of the data under a specific model. These methods include maximum-likelihood estimation (MLE) , expectation-maximization (EM), and Markov chain Monte Carlo (MCMC) methods (Little and Rubin, 2002; Madras, 2002; Kuroda, 2004). These methods offer great flexibility, as the imputation procedures can include models of the covariance structure of the data themselves. Also, the underlying model assumptions can be assessed, and the model subsequently improved, based on standard statistical diagnostics. There are methods for continuous values, categories, and censored data. We present the EM approach to imputation in Section 4.2.4.

4.2.3 Nearest-Neighbor Imputation

Nearest-neighbor methods for imputation of attributes in forest inventories are common in forest resource analysis. The method and various forms of it have been used to impute stand-level attributes (Moeur and Stage, 1995), plot-level attributes (Holmström et al., 2003), basal-area diameter distributions (Haara et al., 1997), and tree lists (Korhonen and Kangas, 1997).

Numerous R packages provide nearest-neighbor methods for continuous and categorical data; for example: the class (Venables and Ripley, 2002) and knncat packages (Buttrey, 2008); weighted k-nn, for example the kknn package (Schliep and Hechenbichler, 2009); and generic nearest-neighbor methods for missing observations, for example the class package (Venables and Ripley, 2002). Crookston and Finley's yaImpute package includes many methods that are commonly found in the forest inventory analysis, such as the gradient nearest-neighbor (GNN, Ohmann and Gregory, 2002), Most Similar Neighbor (MSN, Moeur and Stage, 1995), and the random forest method Breiman (2001a).

Here we will use the simplest of the nearest-neighbor methods, k-nn imputation. The k-nn method is a kernel density estimation method that is commonly used to impute missing values where the value of k dictates the number of nearest neighbors, in statistical distance, to the observed data in any given row of the dataset (Bishop, 2006, Chapter 2). The method is nonparametric in the sense that it does not assume any underlying distribution. The lack of a distributional assumption is an advantage, as the distributional form of variables of interest is often influenced by factors such as market conditions (for example, sales, exchanges, and depletions) and disturbance patterns (such as fire, insects, and disease). It is often difficult to find a distributional form that is sufficiently flexible to match these contingencies.

Nearest-neighbor methods require *training* and *target* datasets. The *training* dataset comprises the data for which we have all the necessary attributes. The *target* dataset is the dataset for which we have some, but not all, of the values. The task then is to find which data in the *training* data are closest to the *target* data, using the attributes that are common to both datasets, and assign those values that are missing from the training dataset.

Normally, we would begin by obtaining our training data from those rows in the inventory that contain all of the attributes. Since there is a possibility that there are rows with all nine attributes missing, we shall include the centroid of the stand polygon so that we can assign the geographically, not statistically, closest stand if we have no other attributes to compare. For this example, location is the only variable that is available for every polygon regardless of missingness.

We already know from the MISSING variable how many attributes are missing from each row of the stands object. We select those rows from stands that have no missing values as our training dataset,

```
> centers <- coordinates(stands)
> stands.frame <- as.data.frame(stands)
> stands.frame$x.ctr <- centers[,1]
> stands.frame$y.ctr <- centers[,2]
> stands.frame$MISSING <- rowSums(is.na(stands.frame))
> training.stands <- subset(stands.frame, MISSING == 0)
> nrow(training.stands)
```

[1] 324

The remainder of the rows contain various missing attributes. Assign those rows to a variable named target.stands,

```
> target.stands <- subset(stands.frame, MISSING > 0)
> nrow(target.stands)
```

[1] 186

Recall from Chapter 2 that stands contains 510 polygons and in Section 4.2.1 we partitioned stands into *target* and *training* rows to represent rows with and without missing data, respectively. From the output above, the number of rows with and without missing data are 324 and 186, respectively, and the sum of the two equals the total number of rows in the stands object.

Next, create the training and target objects and include the coordinates of the centroid for each of the sampled stands

```
> cls <- c("x.ctr","y.ctr")
> training <- training.stands[,cls]
> target <- target.stands[,cls]
```

so that the original dataset is now divided into two datasets (**training** and **target**) and the row names represent the factor we wish to assign to a **target** dataset from the **training** dataset.

To confirm our data are properly split, print the first few rows of each to verify that the datasets as different,

```
> head(rownames(training))
```

```
[1] "0"   "4"   "5"   "7"   "13"  "18"
```

```
> head(rownames(target))
```

```
[1] "1" "2" "3" "6" "8" "9"
```

The resulting printout shows that the row names, or row labels, from the **training** object are not in the **target** object and vice versa. Here the labels are numeric but do not correspond with the row number.

Now, using the **knn** function in the class package, classify the **target** stand polygons using the polygons in the **training** object. The result will be the values to replace the missing attributes in **target** using a subset of the rows from the **training** object. Specifically, the assignments here should come from the nearest neighbors, in terms of Euclidean distance, to the polygons with missing attributes (**target**).

```
> cl <- factor(rownames(training))
> knn.res <- knn(training, target, cl, k = 1)
```

The resulting **knn.res** object contains a **levels** attribute, which represents the class factors (*rowlabels*), assigned from **training** for the **target** dataset. For a more complete description of the function, see the documentation for the **knn** function.

Next, extract the row names from the **target** dataset to identify those rows that originally contained missing values.

```
> missing.rows <- as.character(rownames(target))
```

Then, to keep track of which rows are being replaced in the original data, coerce the results of the nearest-neighbor assignment (**knn.res**) into a numeric vector, named **replacement.rows**,

```
> replacement.rows <- as.character(knn.res)
```

so that now we have two vectors that contain the original row identifiers that correspond to rows with missing values (**missing.rows**) and the resulting rows that should be assigned to replace the missing rows (**replacement.rows**) in the target data (**target**).

Now, to facilitate our analysis, we create a copy of the original data frame, which will be used to construct our final data frame object,

```
> stands.knn <- stands.frame
```

Next, replace the subset of the attribute data row (columns 5 though 13) in the new copy by assigning the subset from the row from the target vector from the *k*-nn assignment to the stands.knn data frame object. The reason we are using a subset of the columns is to maintain the geographic data since the first four variables are specific to each stand polygon and are not influenced by the missing data.

Again, use the indexes of the fields rather than the names of the fields themselves to replace the rows that contain missing values from the original data that does not contain missing values by using the vectors of row identifiers for the respective data frame objects,

```
> stands.knn[missing.rows,c(4,5:13)] <-
+   stands.frame[replacement.rows,c(4,5:13)]
> show.cols.with.na(stands.knn)
```

```
No missing values.
```

To make sure that we can easily keep track of the values that were imputed, we add a replaced.by field that identifies the rows that have replaced the originally missing rows (missing.rows),

```
> stands.knn$replaced.by <- NA
> stands.knn[missing.rows,]$replaced.by <- replacement.rows
```

Finally, check the results by examining the values for the first few stands in the newly created object,

```
> head(as.data.frame(stands.knn))
```

	SP_ID	AREA	PERIMETER	STANDID	ALLOCATION	TAGE	BHAGE
0	0	612731.1	4827.898	010101	forest	42	37
1	1	2466649.0	13244.230	010108	forest	81	76
2	2	226084.4	2019.828	010101	forest	42	37
3	3	139821.6	1807.653	010113	forest	14	15
4	4	2897093.0	12078.560	010105	forest	40	35
5	5	455932.7	2779.376	010113	forest	14	15

	DF_SITE	TOTHT	CUBVOL_AC	TPA	QMD	BA	x.ctr	y.ctr
0	101	62.2	5201	419.84	9.4	202	1264263	400519.8
1	118	102.3	3133	43.59	17.7	75	1265432	399298.9
2	101	62.2	5201	419.84	9.4	202	1265964	400281.0
3	124	31.3	637	458.39	5.3	69	1263518	400345.6
4	95	59.9	2874	304.57	9.3	144	1263896	399065.8
5	124	31.3	637	458.39	5.3	69	1264166	399966.9

```
  MISSING replaced.by
0       0        <NA>
1       1          13
2       6           0
3       6           5
4       0        <NA>
5       0        <NA>
```

```
> show.cols.with.na(stands.knn)
```

```
replaced.by
        324
```

By printing the first few rows, we can verify that the rows with missing
data have had columns replaced (columns 4 and 5 through 13). For example,
the dataframe row labeled "2" of the original data (stands[3,]) has been
replaced by row 0 based on the Euclidean distance. If we print them out
together,

```
> stands.knn[c(1,3),]
```

```
  SP_ID    AREA PERIMETER STANDID ALLOCATION TAGE BHAGE DF_SITE
0     0 612731.1  4827.898  010101     forest   42    37     101
2     2 226084.4  2019.828  010101     forest   42    37     101
  TOTHT CUBVOL_AC    TPA QMD  BA   x.ctr    y.ctr MISSING
0  62.2      5201 419.84 9.4 202 1264263 400519.8       0
2  62.2      5201 419.84 9.4 202 1265964 400281.0       6
  replaced.by
0        <NA>
2           0
```

we can see that the entries we replaced (columns 4 and 5 through 13), contain
the same values. Again, since the **target** and **training** vectors are the row
labels, not row numbers, caution must be taken when creating scripts that
automatically replace values.

Finally, the show.cols.with.na function reports the number of rows
where the replaced.by attribute is NA,

```
> show.cols.with.na(stands.knn)
```

```
replaced.by
        324
```

which matches the number of rows in the training data.

We now have a data frame object (stands.knn) that contains the original
data from stands, and where there were missing values, we replaced a subset
of the row with the associated columns within that row from the original
data that were from the nearest, in Euclidean distance, from the centroid of

the stand. Ultimately we now have a data frame object that does not contain missing values.

To apply k-nn in this context, we have invoked some useful simplifications:

1. We ignored the fact that some of the attributes were more complete than others.
2. We also ignored the fact that the distribution of the missing attributes might not have been equal, although the frequency of the missing attributes was roughly equal for many of the missing values.
3. We replaced the entire row when at least one attribute was missing, not individual fields based on additional information.

With those simplifications, we have generated imputed values similar to those of our observations, where there were no negative values, no extremely large values, and the distributions of the combined imputed and observed values were similar to the observed values alone. This may not always be the case.

Nearest-neighbor classification methods are useful for determining which category a new observation belongs to, but the methods require that the entire training set be stored, which can lead to excessive computational times (Bishop, 2006, Chapter 2). Another drawback is that the methods rely on selecting and computing a weighted combination of the observations rather than making assumptions of the underlying distributions of the missing and observed data.

Yet another problem with our specific method of using the closest neighbor is that there might be large contrasts in adjacent cover types when adjacency is defined by the Euclidean distance and nothing more. For example, if we had a stand polygon with no measurements, it is possible that the stand was recently harvested and could be surrounded by either much older stands or much younger stands. In either case, since we are not including information on the distribution of the surrounding stands, our imputed values may not reflect the true distribution of the current vegetation.

A better method would allow us to control the classification process independently of the size of the training data while maintaining the ability to estimate complex densities. For that, we turn to the expectation-maximization (EM) algorithm.

4.2.4 Expectation-Maximization Imputation

The Expectation-Maximization (EM) algorithm iteratively computes the expected values for missing observations by repeatedly updating maximum-likelihood parameter estimates and imputing expected values until convergence is achieved. It has been applied to maximum likelihood estimation (Little and Rubin, 2002), latent structure models (McLachlan and Krishnan,

2008), neural network learning (Bishop, 2006, Chapter 9), and Gibbs sampling (Watanabe and Yamaguchi, 2004, Chapter 4). For a more complete presentation, we refer the reader to Dempster et al. (1977). Watanabe and Yamaguchi (2004) and McLachlan and Krishnan (2008) also provide a thorough introduction, related models, and extensions to the algorithm.

Briefly, the algorithm iterates over two steps (Little and Rubin, 2002, see Chapter 8):

1. *E-step*: Find the conditional expectation of the missing data from the observed data and current estimates of the parameters and substitute the expectations back into the missing data.
2. *M-step*: Perform maximum-likelihood estimation of the parameters as if there were no missing data.

Imputation using the EM algorithm can be accomplished using the norm package (Novo and Schafer, 2002) if we are willing to assume that our data are distributed as multivariate normal. To use R for imputing missing values using the EM algorithm, first use the `prelim.norm` function,

```
> cls <- c(5:12)
> sd <- as.matrix(as.data.frame(stands[,cls]))
> psd <- prelim.norm(sd)
```

This function performs preliminary bookkeeping functions (e.g., sorting, centering, and scaling) on the input data, and the output object (`psd`) is an intermediate object, so for brevity we do not present the details here. For more complete details, see the documentation provided with the package.

Next, call the `em.norm` function to compute the maximum-likelihood estimates on the `psd` matrix and assign the results to a variable called `thetahat`,

```
> thetahat <- em.norm(psd, showits=FALSE)   #compute mle
```

where `thetahat` contains a vector that represents the maximum-likelihood estimates of the normal parameters that are not in the original scales.

To extract the estimates, covariances, and correlations from `thetahat` in the original scales, call the `getparam.norm` function,

```
> em.params <- getparam.norm(psd, thetahat, corr=TRUE)
```

which extracts the information of interest. Here we extract only the mean (`em.params$mu`),

```
> names(em.params)

[1] "mu"  "sdv"  "r"

> em.params$mu

[1]    0.00000    58.05591    53.21879  115.85964   69.26065
[6] 5890.92943   241.54077    12.42562
```

Finally, to obtain our imputed values, we seed the random number generator,

```
> rngseed(1234567)   #set random number generator seed
```

make a copy of the original data as we did for the stands.knn object,

```
> stands.em <- as.data.frame(stands)
```

and use the imp.norm function to impute the missing values

```
> stands.em[,cls] <- imp.norm(psd, thetahat, sd)
```

using the cls to index the columns to be replaced.

```
> show.cols.with.na(stands.em)

STANDID      BA
      6     101
```

Once again, we now have a complete dataset that contains no missing values. The imp.norm function returns a matrix of the same form as the input matrix (psd) but with all missing values filled in with simulated values drawn from their predictive distribution given the observed data (sd) and the specified parameter (thetahat).

A print of the summaries (from summary(stands.em)) shows a few negative values for fields that are not normally negative (specifically, QMD, TPA, and BA). This reveals that the method can produce values beyond the range of the original data, which in some cases is unacceptable. In the next section, we compare our results for a single metric (site index) and examine some differences between the two methods.

4.2.5 Comparing Results

Comparing results for imputation methods can include numerous metrics (Dempster et al., 1977). Here we compare the results graphically to simplify our presentation. Specifically, we generate a plot of the distributions for the original data and the distributions of the two imputation methods we tried (see Figure 4.3) and discuss them briefly.

First, construct the individual data frame objects,

```
> knn.frame <- data.frame(site = stands.knn$DF_SITE,
+                         method = "KNN")
> em.frame <- data.frame(site = stands.em$DF_SITE,
+                         method = "EM")
> obs.frame <-
+   data.frame(site = subset(stands.frame,
+                            MISSING == 0)$DF_SITE,
+              method = "OBS")
```

Then merge them together using the `rbind` function,

```
> site.frame <- rbind(knn.frame, em.frame, obs.frame)
```

Finally, generate a histogram of the data using the `histogram` function,

```
> histogram(~ as.numeric(site ) | method, data = site.frame,
+              xlab = "Site Index (feet)",
+              type = "density",
+              main = "Stand Site Index Frequency Distributions",
+              breaks=30,
+              layout=c(3,1),
+              index.cond=list(c(3,1,2)),
+              panel = function(x, ...) {
+                 panel.histogram(x, ...)
+                 panel.mathdensity(dmath = dnorm,
+                                   col = "black",
+                                   args = list(mean=mean(x),
+                                     sd=sd(x)))
+              }
+           )
```

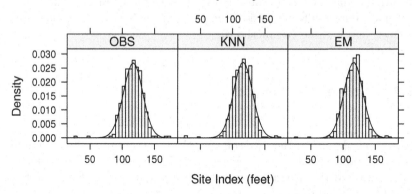

Stand Site Index Frequency Distributions

Fig. 4.3: Distributions of the site index values from the observations (OBS), k-NN, and EM algorithms. An estimated normal density is plotted over the histograms.

Comparing the histograms in Figure 4.3, we can see that the distributions are similar. The ranges between the two sets of distributions are roughly the same, and we are comforted by the fact that in the k-nn imputed distributions, the extreme values are not outside the range of the observations.

The EM algorithm is gaining popularity because of the flexible nature of the method, but packages for estimating non-normal distributions are currently limited. Convergence can be slow, especially when the missing data contain information about the estimated parameters and are functions of the missingness indicator matrix M, as described in Section 4.2.1 (Longford, 1993; Little and Rubin, 2002). While specification of the missing data is at the discretion of the analyst, the sparse dataset can often be viewed as a small portion of a much larger dataset, and within that framework, estimation of the missing values, via the assumed distributions, can be much easier.

The appeal of this method is that a fairly robust method of imputation can be examined with a few lines of code. This still does not free the analyst from examining the assumptions or checking constraints, such as negative estimates or excessively large values.

Although simply assigning attributes from the nearest neighbor yielded acceptable results, several issues were not examined. The fact that abrupt changes in vegetation characteristics often coincide with missing data (e.g., harvest unit not sampled yet), the consequences of ignoring the additional information such as the distributions, or the correlation among variables as well as over distance cannot be determined without more detail. In order to examine the landscape with finer detail, we need to refine our examination to a resolution smaller than the stand polygon. Fortunately, we know the locations of the plots and have a measurement of the site index at several locations. This knowledge could give us the resolution we need. For that resolution, we turn to interpolation.

4.3 Interpolation

In the previous section, we examined methods to impute stand-level data from sampled stand polygons assuming that the attributes of interest were distributed uniformly throughout the stand polygon and that the attributes were normally distributed. Now, we will also estimate values for unsampled locations, but unlike in the previous section, we will interpolate values between point samples to generate a continuous surface. This procedure will require slightly different assumptions.

Following Neteler and Mitasova (2002, Chapter 7), we define *interpolation* as the conversion of a spatially continuous process, such as elevation or site index observations, into a raster representation. There is some debate regarding the spatial continuity of vegetation attributes, but here we assume that the attributes of interest, for example site productivity, which should be considered independent of the vegetation conditions (Newton and Hanson, 1998), are continuous over the region of interest.

4.3.1 Methods of Interpolation

There are numerous methods for interpolating values at unsampled locations, including point interpolation, inverse distance methods, and minimum-variance methods. Point estimation methods can be classified into polygonal and triangular interpolation methods. The relative performance of these methods can vary depending on the estimation criteria (Isaaks and Srivastava, 1989). These methods are relatively simple and are thoroughly presented within other references (Okabe et al., 2000; Cressie, 1993), so we do not present them here.

Instead, we focus on minimum-variance methods. These methods are designed to yield the best estimates under one particular statistical criterion: unbiased estimates with minimum error variance. To begin, we need to address the following questions to provide key contextual information:

- Are we interested in global or local estimates?
- Are we interested in population parameters, such as mean and variance, or do we want the complete distribution of values?
- What is the support? Do we want point values or estimates over larger areas such as polygons?

For global estimates, all of the sample values and locations are used for prediction. Global estimates are typically used for estimating the distribution of attributes over the region of interest and can be greatly influenced by clustering (Isaaks and Srivastava, 1989) . For local estimates, only those points near the point of interest are used for estimation. As with global estimates, local estimates are also influenced by clustering, but local estimation methods account for both the distance of sample points to estimation locations and possible redundancy among samples.

Historically, forest inventories have been focused on global estimation by attempting to answer such questions as "How much volume does a forest contain?" or "How many trees are in a given area?" Site-specific information is also becoming important. Both types of estimation are influenced by the number and location of samples within a given area, and much work has been done to examine the influence of plot location on both global estimation and local estimation (de Gruijter et al., 2006, Chapters 6 and 8 respectively).

Once we have decided whether we are interested in obtaining global or local estimates, we then need to determine whether we are interested in estimating parameters of a distribution or the distribution itself. The mean is the most commonly estimated parameter (Isaaks and Srivastava, 1989). If estimation of an entire distribution is of interest, then both parametric and non-parametric methods are available. Parametric methods make assumptions about the underlying distribution of the data, but the assumptions can be difficult to verify and may not be appropriate for many types of situations where the surface is neither smooth nor continuous. Non-parametric methods do not make those assumptions but require interpolation between points on the cumulative histogram. Should we need to extrapolate beyond the data, the results can be invalid.

Then we need to determine if we are interested in predicting values at individual points across the entire region or in smaller subregions such as stand or harvest polygons rather than individual points. The *support*, or spatial continuity , is important, as it determines the appropriate prediction methods depending on our desire to predict point, line, area, or volume estimates. The support acts as a smoothing filter depending on the size of the area being estimated and the density of the sample data (Goovaerts, 1997, Chapter 5) and (Isaaks and Srivastava, 1989, Chapter 19).

For this section, we will examine one of the most basic types of interpolation commonly used for point support. There are numerous methods that can be used for all types of support, but it is beyond the scope of this book to cover all of the possibilities.

Here we are interested in estimating the mean site index at unsampled point locations over the entire landscape. The data, from Chapter 2, includes the `final.plots` dataframe,

```
> final.plots <- read.csv("../../data/final-plots.csv" )
```

To begin, we examine the distribution of the site index plots using a basic histogram and normal q-q plot,

```
> opar <- par(mfcol = c(1,2), las = 1, cex.axis = 0.70)
> site.plots <- subset(final.plots, !is.na(site))
> hist(site.plots$site, xlim = c(0,220), ylim = c(0,0.025),
+       freq = FALSE, main = "Histogram", xlab = "Site Index",
+       breaks = 30)
> qqnorm(site.plots$site, pch=46)
> qqline(site.plots$site, lty=1, col="grey", pch=46)
```

Judging from the histogram and normal quantile-quantile plots in Figure 4.4, the distribution of the site index is fairly close to normal, which would lead us to conclude that using one of the parametric methods might yield satisfactory results.

Some of the nomenclature associated with the various interpolation methods can cause confusion. Schabenberger and Pierce (2002, Table 9.3) present a useful table of the various features of kriging systems in relation to the underlying assumptions. Here, we need to estimate the site index at any point in a finite region where the mean over the entire region is unknown. Their table suggests that a parametric method for point estimation known as ordinary kriging meets our requirements.

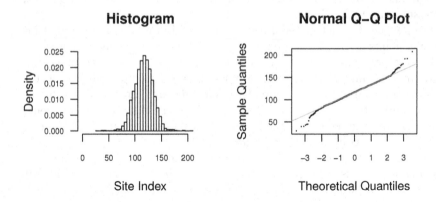

Fig. 4.4: Graphical summaries of the site index observations from the site index plots.

4.3.2 Ordinary Kriging

Ordinary kriging allows the prediction of variables at unmeasured locations by using the relationships among the sample points around the prediction site (Goovaerts, 1997). The method is commonly used when samples are taken from an area and the goal is to obtain complete coverage of an attribute with some measure of uncertainty about the predicted value. Kriging is named after Danie G. Krige, a South African who developed the technique, with Herbert Sichel, in an attempt to more accurately predict ore reserves (Armstrong, 1998), thus the use of terms such as *nugget*.

In order to use this method, we have to make some assumptions about the process we are attempting to predict :

- We must assume the variable we are trying to predict is a continuous random variable over the region of interest.
- We must assume the expected value of the variable is constant over the region.
- We must assume the variance is constant and finite.
- We must assume the covariance function is dependent only on the distance between two points and not the absolute positions of the data.

For an excellent introduction to random variables and random functions within the framework of spatial interpolation, we recommend Isaaks and Srivastava (1989), and for a complete overview of most kriging methods, we recommend Goovaerts (1997) and Schabenberger and Gotway (2005). These references provide an excellent background on spatial interpolation. We provide only a brief introduction here.

4.3.2.1 Theory

Ordinary kriging is a best linear unbiased estimator (BLUE) because the estimates are weighted linear combinations of the data that have the smallest possible variance among all possible linear combinations of the data (Isaaks and Srivastava, 1989). The estimator is unbiased because the mean residual error is zero, and it minimizes the variance of the errors. In order to achieve BLUE, we have to develop a model to describe our data and select weights to apply to our random function model that ensure that the error variance

$$\hat{\sigma}_r^2 = \hat{\sigma}^2 + \sum_{i=1}^{n}\sum_{j=1}^{n} w_i w_j \hat{C}_{ij} - 2\sum_{j=1}^{n} w_i \hat{C}_{i0} \tag{4.1}$$

is minimized subject to the unbiasedness constraint $\sum_{i=1}^{n} w_i = 1$. $\hat{C}$ is a model of the covariance (Isaaks and Srivastava, 1989, Chapter 12). Since we need to constrain the solution to equation (4.1), a system of $n+1$ equations can be expressed more compactly in matrix notation as

$$\begin{bmatrix} \hat{C}_{11} & \dots & \hat{C}_{1n} & 1 \\ \vdots & \ddots & \vdots & \vdots \\ \hat{C}_{n1} & \dots & \hat{C}_{nn} & 1 \\ 1 & \dots & 1 & 0 \end{bmatrix} \begin{bmatrix} w_1 \\ \vdots \\ w_n \\ \mu \end{bmatrix} = \begin{bmatrix} \hat{C}_{10} \\ \vdots \\ \hat{C}_{n0} \\ 1 \end{bmatrix} \tag{4.2}$$

or more simply

$$\mathbf{Cw} = \mathbf{D} \tag{4.3}$$

where $\mathbf{w}$ is a vector of weights, $\mathbf{C}$ is the covariance matrix of the observations, and $\mathbf{D}$ is the vector of the covariances at the points themselves; that is, $\mathbf{C}(h) = 0$, where $\mathbf{h}$ is the distance between sample points. This generates a system of $n+1$ equations that can be easily solved for $\mathbf{C}^{-1}$ to obtain the kriging weights $\mathbf{w}$

$$\mathbf{w} = \mathbf{C}^{-1}\mathbf{D} \tag{4.4}$$

where the resulting values for $\mathbf{w}$ produce the unbiased estimates with the minimum error variance for equation (4.1). Finally, the resulting estimate ($\hat{v}_0$) can then be calculated as

$$\hat{v}_0 = \sum_{i=1}^{n} w_i v_i \tag{4.5}$$

and the resulting minimized estimation variance ($\hat{\sigma}_r^2$) is

$$\hat{\sigma}_r^2 = \hat{\sigma}^2 - \sum_{i=1}^{n} w_i C_{i0} + \mu \tag{4.6}$$

It should be noted here that the use of the system of equations we have presented requires some caveats and observations.

First, since we will need to predict values ($\hat{v}_0$) that are at distances that are not found in the distance matrix, we must construct a function of the distance between points to describe the relationship among the attributes of interest. This relationship $\mathbf{C}(h)$ represents the statistical distances among the sample points separated by a distance h. Thus when two points are very similar, the values of $\hat{\mathbf{C}}_{ij}$ will be large compared with the other entries of the matrix.

Second, there is no guarantee that a $\mathbf{C}^{-1}$ exists that yields a unique solution to our system of equations. We can check for the positive definite condition ($\mathbf{x}'\mathbf{A}\mathbf{x} > 0; \mathbf{A}, \mathbf{x} \in \mathbb{C}^n$) using any number of methods, such as determining whether all of the eigenvalues are greater than zero. In most cases, we do not need to worry about this condition, as we will almost always fit our spatial relationships with positive definite functions, such as the exponential function presented in equation (4.7).

Third, the predictor is an *exact interpolator*, meaning the predictions exactly match the measured values at locations where measurements have been made and the corresponding estimation variance is zero (Journel and Huijbregts, 1978).

Last, the matrix $\mathbf{D}$ is not necessarily the spatial distance from the sample points to the location we are estimating but rather the statistical distance. The combination of these two matrices ($\mathbf{C}$ and $\mathbf{D}$) accounts for the two most important aspects of spatial estimation: clustering of the data values ($\mathbf{C}^{-1}$) and the statistical distances ($\mathbf{D}$) between them (Isaaks and Srivastava, 1989).

In order to compute the kriging weights, we must first decide on a random function, $\mathbf{C}(h)$, that will represent the spatial relationship of our data.

4.3.2.2 Descriptions of Spatial Correlation

In order to obtain the unbiased estimate of the random function $\mathbf{C}(h)$ with the minimum error variance, we need to define the spatial relationship for our random function model. This relationship is usually defined as a model of the covariance structure of the distance between sample locations and data with some unexplainable variation (*nugget*). We use this model to describe the variation of our random function model. The covariance model is often a combination of two models (Armstrong, 1998). For example, when the exponential model $\exp(\frac{-h}{a})$ is combined with the nugget model (c_0), the covariances form the function

$$C(\mathbf{h}) = \begin{cases} c_0 + c_1 & \text{if } \mathbf{h} = 0 \\ c_1 \exp(\frac{-h}{a}) & \text{if } \mathbf{h} > 0 \end{cases} \tag{4.7}$$

which can also be expressed as a semi-variogram

$$\gamma(\mathbf{h}) = \begin{cases} 0 & \text{if } \mathbf{h} = 0 \\ c_0 + c_1(1 - \exp(\frac{\mathbf{h}}{a})) & \text{if } \mathbf{h} > 0 \end{cases} \tag{4.8}$$

where $c_0 + c_1 = \sigma^2$ and h is the distance between two points. The c_0 and c_1 parameters have special interpretations:

1. c_0 describes that part of the variance that cannot be accounted for in the smallest sampling distance between two points. This is known as the *nugget effect*.
2. a is known as the *range*. The *range* is that distance at which the variance is considered constant and the distance between points will not account for the variation. It can be considered the horizon at which no changes in variation can be detected.
3. The sum of $c_0 + c_1$ is known as the *sill*, which represents the variance of the random variables.

In fact, the covariance function is directly related to the semi-variogram in that one can be expressed as a variation of the other (Isaaks and Srivastava, 1989; Goovaerts, 1997; Pebesma, 2004). It can also be shown that the total variance $(\hat{\sigma}_r^2)$ for an ordinary kriging system can be decomposed as

$$\hat{\sigma}_r^2 = \hat{\sigma}^2 + \frac{\hat{\sigma}^2}{n} \tag{4.9}$$

where $\hat{\sigma}^2$ is the variance of the random variables and $\frac{\hat{\sigma}^2}{n}$ is the variance of the unknown mean. Closer examination shows us that the addition of extra samples (increasing n) only reduces the uncertainty about the unknown mean and not the variance introduced by the random variables in our random function.

There are many semi-variogram model forms. The gstat package provides infrastructure to list and examine different forms (Pebesma, 2004). A list of possible semi-variogram models can be obtained using the vgm function with no arguments, and a convenient trellis plot of the semi-variogram models can be generated using the show.vgms function.

Since we wish to predict values and their accompanying confidence intervals, our first task is to develop the covariance (semi-variogram) model for the plot summary data we developed in Chapter 2.

4.3.3 Semi-variogram Estimation

Much has been written on the subject of fitting semi-variograms (Cressie, 1993, Chapter 2). Armstrong (1998) provides three reasons that least-squares methods often fail when fitting semi-variogram models:

1. The function must be positive definite; otherwise mean-squared errors of prediction may be negative.
2. Least-squares fitting assumes all sample points are independent observations.
3. Semi-variogram behavior close to the origin is unknown, yet it is vital, and least-squares methods cannot take this into account.

The gstat package provides several methods to estimate parameters for semi-variogram models and uses iteratively reweighted least squares (Cressie, 1993, Chapter 2) by default (Pebesma, 2004). The package also provides a method for fitting the semi-variogram using restricted maximum-likelihood (REML) estimation, although Pebesma (2004) suggests that REML may be slow for moderate to large datasets (that is, having more than 100 observations). We will use the default settings.

To estimate the semi-variogram for our site index data, we first obtain the plots that have a site index observation by constructing a data frame object that contains the non-missing site index values.

```
> site.plots <- subset(final.plots, !is.na(site))
```

We then build a model of the semi-variogram using the `variogram` and `fit.variogram` functions and fit.

```
> site.var <- gstat::variogram(site ~ 1,
+                               locations = ~x+y,
+                               data = site.plots,
+                               width = 50,
+                               cutoff = 3000)
> site.model <-
+   fit.variogram(site.var,
+               vgm(1000, "Exp", 1000, nugget = 0))
```

Here we have used the scope operator (::) to let R know that we wish to use the `variogram` function provided by the gstat package and not the one provided by the spatial package (Venables and Ripley, 2002).

We then print the results of our model:

```
> site.model
```

```
  model     psill     range
1   Nug  129.1209    0.0000
2   Exp  172.6557  952.9546
```

The sum-of-squares value for the fitted model can be obtained using the `attributes` function.

```
> attributes(site.model)$SSErr
```

```
[1] 14.51226
```

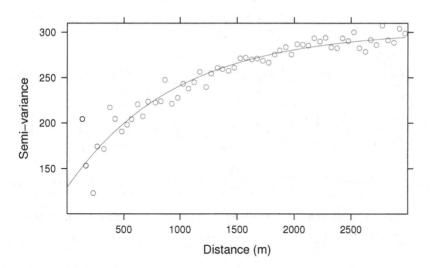

Fig. 4.5: Sample and fitted semi-variograms for the site index.

A plot of the resulting experimental (points) and fitted (line) semi-variograms can be generated using the **plot** function (see Figure 4.5).

```
> plot(site.var,
+       model = site.model,
+       main="Site Index Semi-variogram",
+       ylab = "Semi-variance", xlab = "Distance (m)",
+       ylim = c(100,310), xlim = c(0,3000))
```

The sum of the sills $c_0 + c_1$ for the two individual models is roughly where the semi-variogram approaches a constant value and range. This value at which about 95% of the variance is accounted for occurs near the range of the exponential model (Isaaks and Srivastava, 1989).

```
> sum(site.model[,2])
```

```
[1] 301.7766
```

Before we predict the site index values throughout the landscape, we should also discuss the **width** and **cutoff** arguments to the **variogram** function from the gstat package and describe their influence on the semi-variogram. The **width** argument changes the width of the bin size for h so that the larger the width, the more smoothing occurs. The **cutoff** argument determines the maximum distance at which points are no longer included

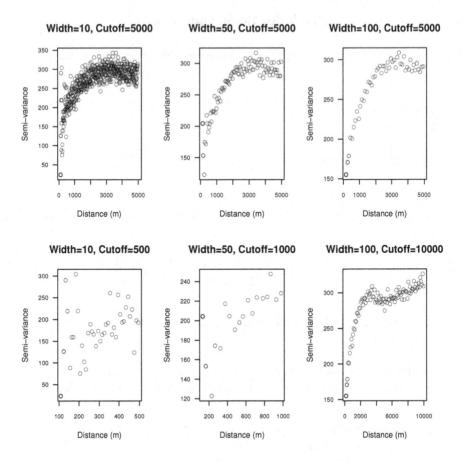

Fig. 4.6: Site index experimental semi-variograms for various widths and cutoffs.

in the semi-variogram model. Figure 4.6 displays several plots with various width and cutoff values.

From Figure 4.6, we can see that selecting various width and cutoff values can influence the parameter estimates and model form of the semi-variogram. Many authors have recommended attempting several width and cutoff values appropriate to the sampling design and caution against a single fit (Isaaks and Srivastava, 1989; Goovaerts, 1997; Armstrong, 1998; Webster and Oliver, 2001). We concur.

4.3.4 Prediction

Upon visual inspection, our semi-variogram model seems to fit the data satisfactorily. Now we can estimate the site index and accompanying variance estimates throughout the entire forest, but first we need to generate a prediction grid and clip the points outside our region using the `inside.owin` function.

First, read the geometry from the boundary file (`boundary.shp`),

```
> bdry2 <- readShapePoly("../../data/boundary.shp")
```

Then, using the `fortify` function in the `ggplot2` package, we generate a region to define the forest boundary using the boundary attribute (`FORBNDRY_`),

```
> dum <- fortify(bdry2, region = "FORBNDRY_")
> summary(dum)
```

```
      long               lat                order
Min.    :1257463   Min.    :354303   Min.    :    1.0
1st Qu.:1267739    1st Qu.:368686    1st Qu.:  345.5
Median :1272965    Median :379267    Median :  690.0
Mean    :1274130   Mean    :379745   Mean    :  690.0
3rd Qu.:1280129    3rd Qu.:391242    3rd Qu.:1034.5
Max.    :1292129   Max.    :400783   Max.    :1379.0
    hole         piece      group            id
Mode :logical   1:1329   2.1:  50   Length:1379
FALSE:1329      2:  50   3.1:1279   Class :character
TRUE :50                 3.2:  50   Mode  :character
NA's :0
```

From the output of the summary function, we have three polygons (`group`) defining the forest boundary: two that define the boundary going clockwise and one to define the hole in the southern portion of the forest.

To generate our boundary, we must manually extract the polygons' pieces, using array indices and our familiarity with the data, and generate a list of points from the inner and outer pieces that can then be used to define an `owin` object, using the poly constructor argument (`owin(poly)`)

```
> dum2 <- dum[dum$id == 3,]
> outer <- dum2[dum2$piece == 1,]
> inner <- dum2[dum2$piece == 2,]
```

Next, construct a list object that contains the pairs of vertices for the boundary

```
> bdry.poly <- vector(2, mode="list")
> bdry.poly[[1]] <-
```

```
+    list(x = outer[(nrow(outer)-1):1,]$long,
+         y = outer[(nrow(outer)-1):1,]$lat)
> bdry.poly[[2]] <-
+    list(x = inner[(nrow(inner)-1):1,]$long,
+         y = inner[(nrow(inner)-1):1,]$lat)
```

and then generate the `owin` object, using the forest polygon boundary we just created,

```
> bdry.owin <- owin(poly=bdry.poly)
```

```
Checking 2 polygons...1, [Checking polygon with 1278 edges...]
2.
done.
Checking for cross-intersection between 2 polygons...1.
done.
```

```
> grid <- gridcentres(bdry.owin, 200, 200)
```

The `owin` function requires that the polygon that defines the window that will be applied masks out unwanted points in our prediction grid. The indices are reversed because the vertices for the outer boundaries must be listed counter-clockwise. The vertices for the holes must be listed clockwise. The object `grid` now contains a prediction grid bounded by the bounding box of the window defined by `bdry.owin`.

Next, reduce the prediction grid by only including those points within the polygonal boundary, defined by the `bdry.owin` object,

```
> ok <- inside.owin(grid$x, grid$y, bdry.owin)
> pred.grid <- data.frame(x=grid$x[ok], y=grid$y[ok])
```

Then, to predict the site index estimates at the new prediction locations using the model we developed previously,

```
> sample.site.plots <- sample(1:nrow(site.plots), 1000 )
> site.pred <- krige(formula = site ~ 1,
+                     locations = ~ x + y,
+                     data = site.plots[sample.site.plots,],
+                     model = site.model,
+                     newdata = pred.grid)
```

[using ordinary kriging]

We can see from printing the first few rows of the resulting predictions (`site.pred`) that our values appear in the range of the distribution presented in Figure 4.7.

```
> head(site.pred)
```

```
       x          y var1.pred var1.var
1 1262056 354419.3   116.2973 236.3804
2 1262229 354419.3   115.8959 242.1147
3 1262403 354419.3   115.3195 246.6632
4 1262576 354419.3   114.8777 248.7282
5 1262749 354419.3   114.5954 247.5300
6 1262923 354419.3   114.3469 242.7019
```

The variable `site.pred` now contains the prediction and the kriging variance at each of the points defined by the prediction grid.

As a simple check of our results, we plot the histogram and normal
quantile-quantile plots as we did with the observations. If we examine the
histograms of the predicted site index values (Figure 4.7) versus the sampled
site index values (see Figure 4.4), we see that the predicted values are well
within the range of the sample values.

```
> opar <- par(mfcol = c(1,2), las = 1, cex.axis = 0.80)
> hist(site.pred$var1.pred, breaks=30,
+       main="Predicted Site Index",
+       freq=FALSE, ylim=c(0,0.08), xlim=c(80,150))
> qqnorm(site.pred$var1.pred, pch=46)
> qqline(site.pred$var1.pred, lty=1, col="darkgrey", pch=46)
```

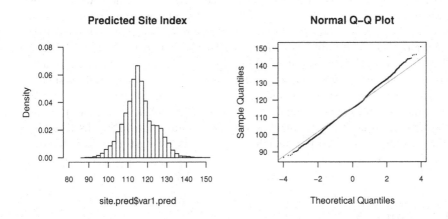

Fig. 4.7: Histogram and q-q plot of the site index plots and the predicted site index
values.

We can see by examining Figure 4.7 that the distribution of the predicted site index values is more peaked than the observations (Figure 4.4)
and has a smaller variance. For now, this is acceptable for our purposes, but

we may want to revisit our assumptions should we wish to apply more stringent assumptions such as matching the distribution of the site index values
throughout the region, but more complex methods (e.g., indicator kriging or
non-parametric methods) are beyond the scope of this presentation.

Finally, generate a plot of the resulting predictions using the following
code (Figure 4.8).

```
> levelplot(var1.pred ~ x + y,
+           data = site.pred,
+           col.regions = terrain.colors(80),
+           main = "Predicted Site Index",
+           xlab = "Easting",
+           ylab = "Northing",
+           panel = function(...) {
+             panel.levelplot(...)
+             lpoints(site.plots[sample.site.plots,]$x,
+                     site.plots[sample.site.plots,]$y,
+                     col="black", cex=0.25, pch=19)
+           }
+         )
```

and the resulting variance estimates similarly (Figure 4.9), both using the
levelplot function in the lattice package.

Examination of Figures 4.8 and 4.9 shows that where we have high con-
centrations of plots, the estimated variance is low, and where we have few
plots, the variance is high. This would suggest that having a well-distributed
grid would yield a good sample that would give a consistent estimate of the
precision of our estimated site index (Webster and Oliver, 2001).

So far, we have only examined estimating a single variable throughout a region. In many cases, we are interested in estimating multiple variables such as
tree density, size, or species composition, which are most likely spatially correlated. As mentioned previously, there are many excellent references presenting techniques for kriging correlated variables, variables in blocks (stands),
and parametric and non-parametric distributions. The gstat package handles
many of these situations.

4.4 Summary

In this chapter, we examined missingness in forest data, a few imputation
methods for estimating missing values, and some of the properties of each
estimation method. We then presented a basic interpolation method (kriging)
to predict values and an associated uncertainty measure (prediction variance)
at unsampled locations over an entire region. The methods we have presented
here constitute only a small portion of possible methods available, so we

Predicted Site Index

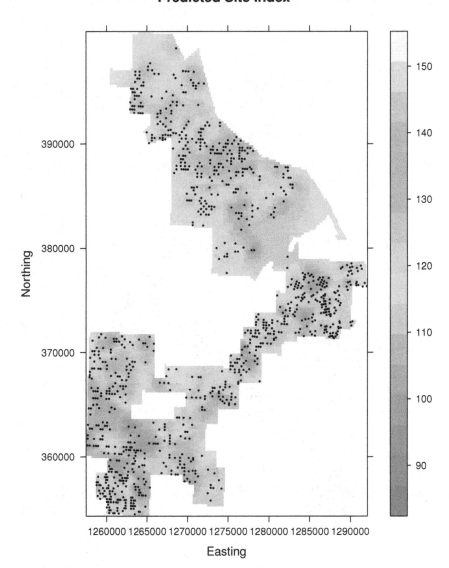

Fig. 4.8: Site index predicted from ordinary kriging of site index plots. The points represent plot locations.

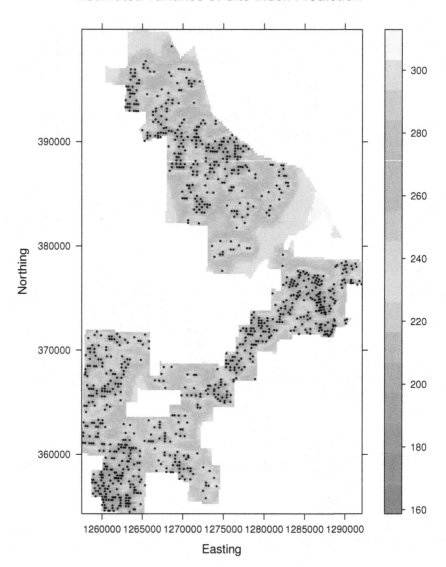

Fig. 4.9: Predicted variance of site index estimates from ordinary kriging of site index plots. The points represent a subsample of the plot locations.

encourage the reader to search the documentation for additional methods that might be more appropriate for any given situation.

Part III
Allometry and Fitting Models

Chapter 5
Fitting Dimensional Distributions

This chapter focuses on the use of statistical tools for fitting models to dimensional data that represent a sample of trees. Here, we consider the models, theory, and tools for one-dimensional data, such as diameter distributions. Two-dimensional data, such as the classical allometric relationships, and multi-dimensional data, which require systems of equations, are handled in a later chapter.

Our goal is to reduce the information about a sample of n observations to as compact and simple a representation as possible while retaining as much information about the underlying population as we can.

The representation of the sample that we choose may require the assumption of a particular model form, for example the Weibull or beta probability density functions, or a family of model forms, such as the Johnson family of distribution functions. Using a model provides several benefits. A model simplifies communication of the outcome because the outcome can be summarized by the estimates of the parameters of the model. Also, if comparable model forms are used, the parametric approach facilitates straightforward connection or comparison with historical theory or results.

The analyst may prefer to represent the data without relying on having to choose a particular functional form, instead using one of a number of non-parametric representations. These so-called non-parametric approaches are more flexible and allow for a wider range of possible densities, including, for example, densities that are bimodal.

It is important to recognize that, since the 1980s, the major differences between parametric and non-parametric models have been in ease of communication and portability. Given an appropriate computing platform, it is just as easy to obtain a prediction of a non-parametric distribution at any number of points as it is a parametric distribution (see, e.g., Borders et al., 1987).

We focus on diameter distributions, but the principles still hold and the tools work for the distribution of any other measures.

5.1 Diameter Distribution

A diameter distribution is a histogram of the tree-level dbh measurements. The bin widths and bin starting point that are used to construct the histogram are chosen subjectively and are usually chosen to simplify the interpretation rather than to optimally capture the shape of the distribution. It is well known that the estimate of the shape of the diameter distribution can be strongly affected by the choice of these parameters (see, e.g., Scott, 1985). In forestry, the width and starting point are most commonly set by convention; for example, a popular configuration is bins that have a width of 5 cm (alternatively, 2 inches), starting at 0 cm.

Diameter distributions are important tools in forest resource management. The interpretation of diameter distributions will affect silvicultural decisions, such as when to thin and how much to thin, as well as harvesting decisions, such as where and when to harvest and what kinds of equipment will be necessary. Diameter distributions are also used as inputs to growth models and sometimes are the subject of growth modeling themselves. Consequently, information about the diameter distribution for a forest stand as it is and as it may be in the future is very useful for forest management. Clutter et al. (1983), Borders et al. (1987), Vanclay (1994), and Avery and Burkhart (2003) provide useful overviews of the uses and interpretation of diameter distributions in forest management.

Invariably the diameter distribution is computed from a sample of trees, rather than the population. Often the sample of trees is arranged in plots, which may muddy the interpretation of the sample-based diameter distribution as an estimate of the population diameter distribution (García, 1992). Trees might compete against their closest neighbors, which may induce negative correlation in dimensions among the proximate trees. Also, local growth patterns could reflect positive correlation due to micro-site variability. These concerns about representativeness and independence of the sample data are most likely impossible to resolve with the usual forest inventory data and are consequently ignored in the modeling phase. However, interpretation of the outcomes can be softened.

From the point of view of the biometrician, the two primary goals for the estimation of a diameter distribution from a sample are prediction in time and in space, and probably both. The spatial goal is to use the sample diameter distributions to represent the size class densities of the population from which the sample was taken. The temporal goal is to project the current diameter distribution forward in time, contingent on some management decision. In order to achieve these tasks efficiently, the distribution is summarized by statistics, and the statistics are projected in time and/or space. The statistics may be parameters of a chosen distribution, such as the Weibull, or they may be quantiles of the sample. The projections are often supported by auxiliary information, such as site class, stem density, or age class. Simplified, the sequence of steps is:

1. a sample is collected,
2. the diameter distribution is constructed,
3. the diameter distribution is represented by statistics,
4. the statistics are projected in time and/or space using, for example, linear regression, and
5. the projected diameter distributions are predicted, conditional on the projected statistics.

In this chapter, we focus on the problem of representing the diameter distribution from the sample data. The problem of projecting the estimates in time and space rely on a suite of other tools, some of which are presented in other chapters of the book.

5.2 Non-parametric Representation

Figure 5.1 shows a simulated random sample of 30 tree diameters, selected from the $N(50, 36)$ distribution. The tree diameters have been lumped into 5 cm (2 in.) diameter classes, forming the diameter distribution. We note in passing that the distribution does not appear to be symmetric, although we know that the population is symmetric.

```
> set.seed(1)
> diameters <- rnorm(30, mean = 50, sd = 6)
> hist(diameters, main = "", xlab = "Diameter Class (cm)")
```

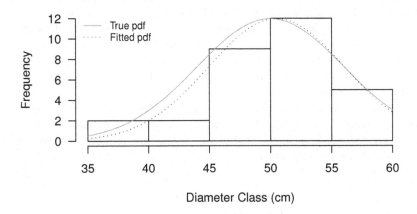

Fig. 5.1: Histogram of 30 sample diameters that were randomly generated from the normal distribution with mean 50 and standard deviation 6, with the true pdf and the pdf of the best-fitting normal curve superimposed.

The sample diameter distribution can be interrogated directly for quantities of interest. For example, the percentile method due to Borders et al. (1987) represents the distribution using a collection of quantiles. The usual empirical quantiles can be obtained from the sample using the `quantile` function, although the 0 and 100% quantiles require the `min` and `max` functions, respectively.

```
> min(diameters)
```

```
[1] 36.7118
```

```
> quantile(diameters, p = (5 + (0:9) * 10) / 100 )
```

```
      5%       15%       25%       35%       45%       55%       65%
39.46411 45.48462 47.39024 49.67330 50.48011 52.32096 53.37517
     75%       85%       95%
54.25220 55.30858 58.02619
```

```
> max(diameters)
```

```
[1] 59.57168
```

5.3 Parametric Representation

Sometimes we prefer to make an assumption about the shape of the underlying density. Here, we will proceed under the assumption that the population is normally distributed. This is a fairly unusual assumption for fitting diameter distributions; we make it here to simplify this introductory material.

The normal distribution is completely determined by two parameters, mean and variance; that is, if you know the mean and variance of data that are normally distributed, then you know everything about the distribution of those data. Therefore, in order to obtain all possible information from a sample about a normally distributed population, all we need to estimate are the mean and variance.

We often prefer to think about the standard deviation rather than the variance because the standard deviation is in the same units as the data, as is the mean. Doing so allows us to think about the spread of the data in units that have a natural interpretation.

5.3.1 Parameter Estimation

We now present a number of approaches to the challenge of estimating the parameters of the distribution given data. We develop these estimators for

the relatively familiar normal pdf before introducing the more popular pdfs
that are used for diameter distribution modeling.

5.3.1.1 Plug-in Principle

One intuitively satisfying way of estimating the mean and standard deviation
is to calculate the sample mean and the sample standard deviation and use
those values to estimate the population parameters. This is called the plug-in
principle (PP),

$$\hat{\mu}_{PP} = \bar{x}; \qquad \hat{\sigma}_{PP} = \sqrt{\frac{1}{n-1}\sum_{i=1}^{n}(x_i - \bar{x})^2}$$

The sample mean is known to be an unbiased estimator of the population
mean. Furthermore, it can be shown that, in terms of estimator variance,
the sample mean is as good as the best possible unbiased estimator for the
population mean; specifically, the variance of the sample mean is the same as
the Cramer–Rao lower bound, which is a theoretical lower limit on how small
the variance of an estimator can possibly be (see, e.g., Casella and Berger,
1990).

The sample standard deviation is a biased estimator of the population
standard deviation. The bias in the estimate can be demonstrated by recog-
nizing that the sample variance is an unbiased estimator of the population
variance and the standard deviation is a curvilinear function of the variance,
and applying Jensen's Inequality (Jensen, 1906).

So, the PP estimators of the mean and standard deviation of the population are:

```
> (x.bar.pp <- mean(diameters))

[1] 50.49475

> (s.x.pp <- sd(diameters))

[1] 5.544725
```

At this point, we have a model of the diameter distribution. The model
form was selected based on an assumption about the distribution of the underlying population and parameterized using a sample of data. We can now
use this model to speculate, for example, what is the most common diameter in the population. Also, if we know how many stems we have in the
population, then we can estimate the number in each size class.

We can superimpose the estimated population distribution over the histogram of the data using the following code (Figure 5.1).

```
> curve(dnorm(x, x.bar.pp, s.x.pp) * length(diameters) * s.x.pp,
+       add = TRUE,
+       col = "darkgrey")
```

Note that we can choose to summarize the sample using its mean and standard deviation but without making an assumption about the underlying distribution. Our purpose here is to push our knowledge one step further so that we can also estimate the relative numbers of stems in the different size classes. Note that we can only push the knowledge further by making assumptions that while they may be plausible, cannot be proven.

5.3.1.2 Method of Moments

A more formal approach to obtaining estimates for the parameters involves equating the sample and population moments. A *moment* is a unit that is used to characterize a random variable that has a certain pdf. Specifically, the k-th moment of a random variable X is the expectation of X^k.

We now demonstrate the method of moments estimation using our data. We have a sample of size n selected from a population that we assume follows the normal distribution, and we wish to estimate the parameters of the normal distribution, μ and σ.

Although the moments are well known for the normal distribution, and indeed most of the distributions that are within the scope of this chapter, we will provide the steps that would be necessary to use for a distribution for which these quantities are unknown. This development draws from Casella and Berger (1990).

In order to compute the moments associated with any pdf, we use the moment generating function (mgf). This function is defined as

$$M_X(t) = E e^{tX} \qquad (5.1)$$

that is, the mgf of the random variable X evaluated at a value t is the expectation of e raised to the power of tX. Then, the moments of X are generated as follows: the k-th moment is is the k-th derivative of $M_X(t)$ with respect to t, evaluated at $t = 0$.

The mgf for the normal distribution is therefore

$$M_X(t) = \frac{1}{\sqrt{2\pi}\sigma} \int_{-\infty}^{\infty} e^{tX} e^{-(x-\mu)^2/(2\sigma^2)} dx \qquad (5.2)$$

and after some algebra,

$$M_X(t) = e^{\mu t + \sigma^2 t^2/2} \qquad (5.3)$$

So, the first moment is

$$\mu_1 = \frac{d}{dt} e^{\mu t + \sigma^2 t^2/2} \big|_{t=0}$$

which is μ. The second moment is

$$\mu_2 = \frac{d^2}{dt^2} e^{\mu t + \sigma^2 t^2 / 2} \big|_{t=0}$$

which is $\mu^2 + \sigma^2$.

Then we set these moments equal to the first two moments of the sample and solve the resulting simultaneous equations. That is, take $m_1 = \frac{1}{n}\sum x$ and $m_2 = \frac{1}{n}\sum x^2$ and set them equal to μ_1 and μ_2, respectively. Then, after further manipulations, $\hat{\mu} = \bar{x}$ and $\hat{\sigma} = \sqrt{\frac{1}{n}\sum (x_i - \bar{x})^2}$. In R,

```
> (x.bar.mm <- mean(diameters))
```

```
[1] 50.49475
```

```
> (s.x.mm <- sqrt(sum((diameters - x.bar.mm)^2) /
+                 length(diameters))))
```

```
[1] 5.45153
```

This several-step process will reliably provide estimates, but there is no body of theory to suggest that method of moments estimates are always particularly good.

5.3.1.3 Maximum Likelihood

Another popular parameter estimation technique involves direct use of the pdf of the assumed underlying population: maximum likelihood. Maximum likelihood involves maximizing the joint probability density function of the data *as a function of the parameters*. That is, the maximum-likelihood estimates (MLE) of the parameters are those values that maximize the joint probability density of the data, conditional on the data and on having chosen a population distribution.

For mathematical convenience and computational stability, it is better to maximize the log of the joint density. This is equivalent to maximizing the sum of the logs of the densities, if the data are independent. In addition to computational stability, another advantage of working with the log of the joint density is that large-sample estimates of the standard errors of the parameters can be easily obtained, as we will see below.

We compute MLE estimates by applying an optimizer to the sum of the logs of the probability density functions.

```
> ll.norm <- function(parameters, data) {
+    mu.hat <- parameters[1]
+    sigma.hat <- parameters[2]
+    return(sum(dnorm(data, mu.hat, sigma.hat, log = TRUE)))
+ }
```

The optimizer requires initial estimates. Initial estimates can be generated using the plug-in principle or the method of moments. We scale the function by -1 using the `fnscale` argument of the control list because `optim` minimizes the function, and we want to find the values of the parameters that maximize the log-likelihood. The default engine for `optim` is the Nelder–Mead simplex algorithm (Nelder and Mead, 1965).

```
> (mle.n <- optim(c(x.bar.mm = x.bar.mm, s.x.mm = s.x.mm),
+               ll.norm,
+               data = diameters,
+               hessian = TRUE,
+               control = list(fnscale = -1)))

$par
x.bar.mm    s.x.mm
50.49475   5.45153

$value
[1] -93.44504

$counts
function gradient
      53       NA

$convergence
[1] 0

$message
NULL

$hessian
            x.bar.mm      s.x.mm
x.bar.mm   -1.009449    0.000000
s.x.mm      0.000000   -2.018899
```

A brief description of the preceding output follows. The `par` slot contains the three parameter estimates in the same order as they are used in the `ll.w3` function. The `value` slot reports the value of the objective function evaluated at its maximum — this is the maximized joint log-likelihood. The `hessian` slot reports the estimated Hessian matrix, which is the matrix of second derivatives of the joint log-likelihood with respect to the three parameters, evaluated numerically at the estimated optimum. The other three slots provide feedback on the optimization process itself: the number of evaluations of the function and its gradient, whether or not the convergence criterion had been achieved (0 means that it had been), and textual feedback from the optimizer.

The ML estimators of the mean and standard deviation of the population are then

```
> mle.n$par
```

```
x.bar.mm   s.x.mm
50.49475   5.45153
```

An asymptotic estimate of the covariance matrix of the parameter estimates can be obtained by inverting the negative of the Hessian matrix. We use the following code:

```
> solve(-mle.n$hessian)
```

```
            x.bar.mm       s.x.mm
x.bar.mm  0.9906392  0.0000000
s.x.mm    0.0000000  0.4953194
```

Furthermore, asymptotic estimates of the standard errors can be found using

```
> sqrt(diag(solve(-mle.n$hessian)))
```

```
 x.bar.mm     s.x.mm
0.9953086  0.7037893
```

Notice that the estimate of the mean is the same as those using PP and MM, within reasonable computation error, but the ML estimate of the standard deviation is smaller than the PP estimate. It turns out that PP, ML, and MM estimates of parameters are often identical. When they differ, it is often easier to find the MLE than to find the PP or MM estimates of parameters.

In general, the statistical properties of MLE are at least as good as the properties of PP and MM estimates, and are known to be better under certain circumstances. Specifically, MLEs are consistent and asymptotically efficient, as long as the support of the likelihood is independent of the parameters being estimated. The latter condition is not satisfied, for example, in the case of the location parameter for the three-parameter Weibull. In any case, the characteristics mentioned provide a rationale for selecting ML estimates in the first instance, but they do not guarantee that ML is the best approach to use in any given situation. Thought and care are still required.

In addition to hand-coding the joint log-likelihood, we can also make use of the flexible fitdistr function from the MASS package, which provides maximum-likelihood estimation for a range of univariate distributions (Venables and Ripley, 2002).

```
> library(MASS)
> d.n <- fitdistr(diameters, "normal")
> d.n
```

```
      mean             sd
 50.4947490      5.4515297
( 0.9953086)   ( 0.7037895)
```

5.3.2 Some Models of Choice

Numerous distribution families have been proposed to match diameter distributions; for example, the exponential distribution, the two- and three-parameter Weibull distributions (Bailey and Dell, 1973), and Johnson's S_B (Hafley and Schreuder, 1977), among many others. Some of these distributions are easy to parameterize; others are more difficult.

We start with the two-parameter Weibull distribution. The Weibull distribution originated in the statistical analysis of failure rates and is popular in the statistical analysis of survival data. The interested reader can learn more from Johnson et al. (1994). The two-parameter Weibull can be fit to a dataset in a number of ways; we use the `fitdistr` function again.

```
> d.w2 <- fitdistr(diameters, "weibull")
> d.w2

     shape           scale
 11.5600440     52.8091906
 ( 1.6694364)  ( 0.8766854)
```

Now we want to compare the two distribution fits; that is, compare the fit of the two-parameter Weibull against the fit of the normal distribution. The pdfs both have the same number of parameters, so a direct comparison of the log-likelihoods will yield the same outcome as a comparison of AIC would.

```
> logLik(d.n)
```

'log Lik.' -93.44504 (df=2)

```
> logLik(d.w2)
```

'log Lik.' -91.75936 (df=2)

These results suggest that the two-parameter Weibull distribution is a slightly better fit to our data than the normal distribution.

The three-parameter Weibull distribution is an extension of the two-parameter Weibull that provides for a shift in location. The three-parameter Weibull distribution is useful when the mass of points is known to be remote from zero, which is often the case with tree dimension measurements. The pdf of the three-parameter Weibull distribution is

$$f_X(x \mid \alpha, \beta, \gamma) = \frac{\gamma}{\beta} \left(\frac{x - \alpha}{\beta} \right)^{\gamma-1} \exp\left(-\left(\frac{x - \alpha}{\beta} \right)^{\gamma} \right) \qquad (5.4)$$

with $\beta > 0$, $\gamma > 0$, and $-\infty < \alpha < \infty$. And, in R, we can use the following function:

```
> dweibull3 <- function(x, gamma, beta, alpha) {
+    (gamma/beta)*((x - alpha)/beta)^(gamma - 1) *
```

```
+        (exp(-((x - alpha)/beta)^gamma))
+ }
```

So the joint log-likelihood for an independent sample from the three-parameter Weibull distribution is, ready for optim,

```
> ll.w3 <- function(p, data)
+        sum(log(dweibull3(data, p[1], p[2], p[3])))
```

Note that we have separated the steps of computing the pdf and taking its log for clarity. Using a function that computed the log of the density directly would be more efficient and more stable for operational uses. Density functions that are provided in R, for example dnorm, offer the option of returning the log of the density value directly. Doing so is recommended when the log of the density is of interest, such as in this case, because it eliminates redundant steps; note, for example, that among other things we are taking the log of the exponential raised to the power $-((x - \alpha)/\beta)^\gamma$.

A challenge in fitting this distribution is that the location parameter α limits the domain of the random variable. Maximum-likelihood inference is problematic for such models because the likelihood is undefined across a set of the real numbers that is determined by the data. Furthermore, estimates of such parameters that arise from ML procedures no longer necessarily have the desirable characteristics of consistency and asymptotic efficiency (Casella and Berger, 1990).

From the point of view of parameter estimation, the troublesome portion of the pdf is $(x - \alpha)^\gamma$, which is

- undefined if $\alpha > x$ *and* γ is not an integer,
- negative if $\alpha > x$ *and* γ is an odd integer, and
- increasing as α increases in absolute value for α outside the range of x if γ is an even integer.

Fortunately, the optimization algorithms are reasonably robust against these difficulties and will recover gracefully if they find themselves arriving in areas of the parameter space for which the function is undefined. The only requirement is that the starting values be chosen in such a way that the function can be evaluated. For example, upon application to our data and the three-parameter Weibull log-likelihood, the Nelder–Mead algorithm reports convergence and provides plausible parameter estimates.

```
> mle.w3.nm <- optim(c(gamma = 1, beta = 5, alpha = 10),
+                    ll.w3,
+                    data = diameters,
+                    hessian = TRUE,
+                    control = list(fnscale = -1))
> mle.w3.nm$par
```

```
    gamma        beta       alpha
 6.843394  32.530438  20.122516
```

The log-likelihood, evaluated at its maximum, is

```
> mle.w3.nm$value
```

```
[1] -91.99055
```

We can obtain asymptotic estimates of the standard errors from the result
using code similar to that which we have used previously.

```
> sqrt(diag(solve(-mle.w3.nm$hessian)))
```

```
    gamma        beta       alpha
 3.796482  16.293066  16.079340
```

Different optimization engines may well yield different estimates of the
parameters. For example, the optimization method due to Broyden, Fletcher,
Goldfarb, and Shanno (see, e.g., Nocedal and Wright, 2006) fails to converge
after (the default) 100 iterations, and when more iterations are requested,
using the `maxit` option in the `control` argument, the solution is different
from that offered by Nelder–Mead.

```
> mle.w3.bfgs <- optim( c(gamma = 1, beta = 5, alpha = 10),
+                       ll.w3,
+                       method = "BFGS",
+                       data = diameters,
+                       hessian = TRUE,
+                       control = list(fnscale = -1,
+                                      maxit = 1000))
> mle.w3.bfgs$value
```

```
[1] -91.75805
```

```
> mle.w3.bfgs$par
```

```
      gamma        beta        alpha
 11.6698254  53.2463892  -0.4232776
```

```
> (mle.w3.bfgs.se <- sqrt(diag(solve(-mle.w3.bfgs$hessian))))
```

```
    gamma        beta       alpha
 13.15200  57.46787  57.22259
```

These results leave the analyst in somewhat of a dilemma. The outcome of
fitting the pdf seems to depend on the optimization engine that has been chosen. And, if the parameter estimates from BFGS are used as a starting point
for the Nelder–Mead, then different estimates again are obtained (results not
shown here). However, reference to the estimated standard errors suggests

that the differences between the parameter estimates are minor compared
with the uncertainties in the estimates themselves. Another way to obtain
insight into such differences is to plot the functions and see how substantial
the differences between the fitted curves are (Figure 5.2). The figure suggests
that the practical differences between the fits are negligible.

```
> hist(diameters, main="", xlab="Diam. Class (cm)", freq=FALSE)
> curve(dnorm(x, x.bar, s.x), add=TRUE, col=grey(0.2), lty=1)
> curve(dweibull(x, d.w2$estimate[1], d.w2$estimate[2]),
+         add=TRUE, col=grey(0.2), lty = 2)
> w3.n.h <- mle.w3.nm$par
> curve(dweibull(x - w3.n.h[3], w3.n.h[1], w3.n.h[2]),
+         add=TRUE, col=grey(0.2), lty = 3)
> w3.b.h <- mle.w3.bfgs$par
> curve(dweibull(x - w3.b.h[3], w3.b.h[1], w3.b.h[2]),
+         add=TRUE, col=grey(0.2), lty = 4)
> legend("topleft", lty = 1:4, bty = "n", cex=0.8,
+         legend = c("Normal","Weibull (2)",
+            "W. (3, NM)","W. (3, BFGS)"))
```

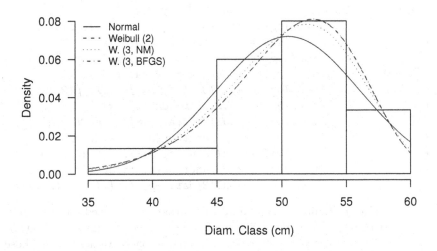

Fig. 5.2: Comparison of normal, two-parameter Weibull, and three-parameter
Weibull fitted curves to the example diameter dataset. The Weibull 2 and the Weibull
3 BFGS are indistinguishable.

5.3.3 Profiling

The behavior of a likelihood depends on the data to which it is fitted and should be the subject of further examination. One way to examine the behavior is to develop and graph the log-likelihood profile curves. To create a one-dimensional log-likelihood profile curve for a given parameter, we first select a set of candidate values for that parameter. For each candidate value we fix the parameter at that value and maximize the log-likelihood across the other parameters. We then plot the maximized log-likelihood values against the candidate values. Analogous graphics can be created for two-dimensional profiles.

Here we develop the code required to provide a reasonably general solution to the problem of creating one- and two-dimensional profile plots for the three-parameter Weibull likelihood. First we create a suite of plot wrappers that allow us to choose which parameters to fix and which ones to vary in our calls to the `dweibull` function. For example, for the location profile, we write the `loc` function, which passes shape and scale as the first argument, which `optim` uses for its optimization, and location as the third argument, which will be considered fixed and will be taken from the suite of candidate values that we specify ahead of the exercise.

```
> all.w3 <- list(loc = function(p, data, fix)
+                  sum(log(dweibull3(data, p[1], p[2], fix))),
+                    sha = function(p, data, fix)
+                  sum(log(dweibull3(data, fix, p[1], p[2]))),
+                      sca = function(p, data, fix)
+                  sum(log(dweibull3(data, p[1], fix, p[2]))),
+                    loc.sha = function(p, data, fix)
+                  sum(log(dweibull3(data, fix[1], p, fix[2]))),
+                      sca.sha = function(p, data, fix)
+                  sum(log(dweibull3(data, fix[1], fix[2], p))),
+                      sca.loc = function(p, data, fix)
+                  sum(log(dweibull3(data, p, fix[1], fix[2]))))
```

We now select the suites of candidate values that will be fixed for the conditional optimization. We arbitrarily nominate the range of values as being the estimated standard errors and choose 30 equally spaced values within those extremes as follows.

```
> grain <- 30
> k <- 1
> p.hat <- mle.w3.bfgs$par
> p.se <- mle.w3.bfgs.se
> frame <- list(sha = seq(from = p.hat[1] - k * p.se[1],
+                          to = p.hat[1] + k * p.se[1],
+                          length = grain ),
```

```
+                   sca = seq(from = p.hat[2] - k * p.se[2],
+                        to = p.hat[2] + k * p.se[2],
+                        length = grain ),
+                   loc = seq(from = p.hat[3] - k * p.se[3],
+                        to = p.hat[3] + k * p.se[3],
+                        length = grain ))
```

We now want to simplify the conditional maximization of the log-likelihood across the suite of candidate values. We rewrite the call to optim and wrap it in a function that allows us to pass the fixed values as an argument, one by one, as well as the version of the likelihood wrapper that we wish to use and an integer to identify which starting parameter estimates to retain. We use the parameter estimates from the previous fit as the starting points to provide some stability. We wrap the call to optim in the try function because there is no guarantee that our arbitrarily selected fixed parameter values will result in an optimizable function, and we want to handle the ensuing errors gracefully.

```
> profile.fn <- function(x, fn, p.fix) {
+    out <- try(optim( p.hat[-p.fix],
+                      fn,
+                      method = "BFGS",
+                      data = diameters,
+                      fix = x,
+                      control = list(fnscale = -1,
+                          maxit = 1000)), silent = TRUE)
+    if (class(out) == "try-error") return(NA)
+    else return(out$value)
+ }
```

We can then call this function in sapply. We can now compute the profile curves and plot them (Figure 5.3).

```
> sha.out <-
+    sapply(frame$sha, profile.fn, fn = all.w3$sha, p.fix = 1)
> sca.out <-
+    sapply(frame$sca, profile.fn, fn = all.w3$sca, p.fix = 2)
> loc.out <-
+    sapply(frame$loc, profile.fn, fn = all.w3$loc, p.fix = 3)
> ll.stack <- as.data.frame(rbind(cbind(sha.out, frame$sha),
+                                 cbind(sca.out, frame$sca),
+                                 cbind(loc.out, frame$loc)))
> names(ll.stack) <- c("ll","x")
> ll.stack$parameter <- rep(c("shape","scale","location"),
+                            rep(grain,3))
```

Finally, plot the profiles using xyplot from the lattice package.

```
> xyplot(ll ~ x | parameter,
+                 type = "l",
+                 ylab = "ll(x)",
+                 scales = list(x="free"),
+                 layout = c(3,1),
+                 data = ll.stack)
```

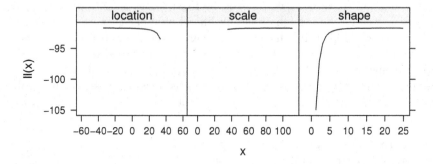

Fig. 5.3: One-dimensional profile plots for a three-parameter Weibull function fitted to the 30 normally distributed, randomly generated diameter values.

These profile curves look quite poor. We would like to see inverted parabolas with the apex at the MLE. These graphs suggest that our efforts to fit the three-parameter Weibull to the sample data do not necessarily yield a model that can be extrapolated reliably. The profiles are far from symmetric about the estimate, and indeed in some cases do not identify the same optimum that the optimization procedure did. We speculate that the smallness of the sample size may well play a role in this difficulty. It is an instructive exercise to repeat the process with many more simulated observations — say 300 — and see what effect a larger sample has on the model-fitting process. We leave this exercise for the reader.

For higher-dimensional functions, it is also useful to examine the two-dimensional profile plots to see how the information about pairs of parameter estimates varies as a function of each of them. As above, this is a relatively easy problem to set up. Taking the scale and shape parameters as a starting point, we create a two-dimensional grid of values using the expand.grid function.

```
> sha.sca.grid <- expand.grid(sha = frame$sha, sca = frame$sca)
```

We then use a slightly altered version of the call to sapply as the core of a call to mapply, a multivariate equivalent.

```
> sha.sca.grid$ll <-
+   with(sha.sca.grid,
+       mapply(function(sha, sca) {
+         out <- try(optim(p.hat[3],
+                           all.w3$sca.sha,
+                           method = "BFGS",
+                           data = diameters,
+                           fix = c(sha, sca),
+                           control = list(fnscale = -1,
+                             maxit = 1000)))
+         if (class(out) == "try-error") return(NA)
+         else return(out$value)
+       }, sha, sca))
```

Finally, we remove the combinations of values that resulted in a failure to fit.

```
> sha.sca.grid <- na.omit(sha.sca.grid)
```

The results are ready to present. We opt for a contour plot of the join log-likelihood against the two parameters, using the `contourplot` function from the `lattice` library. This function allows us to augment the contours with points that represent the places where the model fit converged, providing feedback on which portions of the parameter space were available.

```
> contourplot(ll ~ sha * sca, data = sha.sca.grid,
+             at = c(-92, -96,-100, -110),
+             xlab = "Shape", ylab = "Scale",
+             panel = function(x, y, z, ...) {
+               panel.contourplot(x, y, z, ...)
+               panel.points(x, y, col="darkgrey")
+               panel.points(p.hat[1], p.hat[2],
+                         pch = 19, col = "black")
+             })
```

The outcome is Figure 5.4. This plot shows that a large proportion of the sample space led to convergence failure for our chosen starting points. We may have been able to obtain better results with a more thoughtful approach. The parameter estimates are highly correlated, and the data do not provide much assurance about their true values.

Again, it is an instructive exercise to repeat the process with many more simulated observations — say 300 — and see what effect a larger sample has on the model-fitting process. We leave this exercise for the reader.

This example suggests that the blithe application of maximum-likelihood estimation to data may result in estimates that have poor properties, especially when sample sizes are small. Caution is recommended, and R provides suitable tools for learning more about the circumstances.

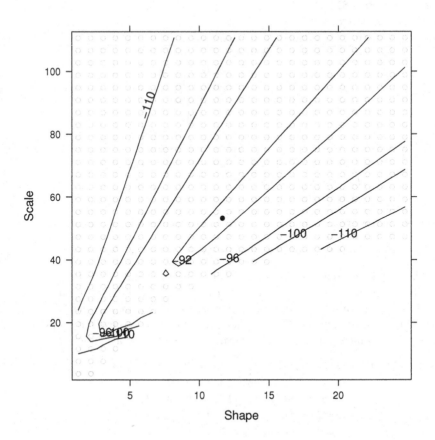

Fig. 5.4: Two-dimensional parameter profile plot for a three-parameter Weibull function fitted to the 30 normally distributed, randomly generated diameter values.

5.3.4 Sampling Weights

Often in forest inventory the sample trees are selected using variable-radius plot sampling, which leads to their sample weights being different from one another; specifically, the sampling weights are approximately proportional to the tree's basal area. In order to estimate the parameters of the distribution appropriately, we need to incorporate the sampling weight into our parameterization of the distribution. Gove (2003) provides a useful discussion, as well as MM and ML estimators for the parameters of the two- and three-parameter size-biased Weibull pdf, where sampling may be proportional to length or area. Here we focus on sampling proportional to area. A simpler correction is used for length-biased sampling; see Gove (2003) for further details.

The alterations that are necessary to make the joint log-likelihood suitable for area-biased estimation are as follows. We compute a correction to the log-likelihood that allows for the fact that the observations have been drawn with variable probability. This correction is the following function of the parameters for area-biased sampling (equation (2), Gove, 2003):

$$\mu_2' = \beta^2 \Gamma(2/\gamma + 1) + 2\alpha\beta\Gamma(1/\gamma + 1) + \alpha^2 \tag{5.5}$$

```
> ll.w3.sb2 <- function(p, data) {
+    gam <- p[1]
+    bet <- p[2]
+    alp <- p[3]
+    mu.prime.2 <- bet^2 * gamma(2/gam + 1) +
+      2 * bet * gamma(1/gam + 1) * alp + alp^2
+    sum(log(dweibull3(data, gam, bet, alp))) -
+      log(mu.prime.2) * length(data)
+ }
```

We can now use `optim` to maximize the joint log-likelihood as we have before, and can furthermore obtain asymptotic estimates of the standard errors of the parameter estimates.

```
> mle.w3.sb2 <- optim( c(gamma = 1, beta = 5, alpha = 0),
+                      ll.w3.sb2,
+                      method = "BFGS",
+                      data = diameters,
+                      hessian = TRUE,
+                      control = list(fnscale = -1,
+                                     maxit = 1000))
```

The function converges, providing the following estimates:

```
> mle.w3.sb2$par

    gamma      beta     alpha
 13.67537  68.21522 -16.45170
```

```
> sqrt(diag(solve(-mle.w3.sb2$hessian)))

    gamma      beta     alpha
 20.10398 100.20780 100.04163
```

The estimated standard errors for the parameter estimates are quite large relative to the estimates themselves. The overarching message from the statistics would appear to be that fitting a three-parameter Weibull function to these diameters is a fairly perilous prospect.

Chapter 6
Linear and Non-linear Modeling

This chapter describes some of the tools that are available in R for fitting certain kinds of conditional distributions; that is, constructing models to predict the behavior of one random variable given that the value of another one or more is known. Examples of such models in forestry include height–diameter models, diameter–volume models, and so on. Such models are of interest for two reasons: in order to make predictions and in order to estimate and interpret the parameters that describe the relationship. For example, a scientist might wish to know whether or not coring trees affects their growth and mortality, and how much; this problem is more naturally an interpretation and estimation problem. Alternatively, a manager might wish to predict heights for some trees for which only diameters and species are known; this problem is a prediction problem. The intended application of the model intimately affects the fitting process. Breiman (2001b) and the discussions that follow are excellent reading on this topic.

6.1 Linear Regression

There are many excellent references that describe the theory and practice of linear regression. Here we will provide enough information to motivate our example and provide context for the analysis. For further reading, we suggest Draper and Smith (1998), Harrell (2001), Weisberg (2005), and Gelman and Hill (2007).

Broadly speaking, a goal of regression is to make a statement about the distribution of the response variable Y conditional on the values of predictor variable(s) X having some known value(s), $F(Y \mid X_1, \ldots, X_p)$.

Usually we are interested only in the conditional mean and the conditional variance, and we ignore the higher-order moments such as the skew. Evidence of residual skew or kurtosis in the data, conditional on the model, is taken as a sign of poor model choice, and ameliorative action is usually taken, such as deploying a generalized linear model or making a transformation of the response variable.

Furthermore, in performing, for example, ordinary least-squares linear regression, we will assume that the conditional distribution of Y given X is normal. The mean and variance are (jointly) a sufficient statistic for the normal distribution, and other moments are known and fixed. If this assumption is untenable, then some alternative approach must be found, which we will touch upon later.

So, we can simplify the expression of our model as

$$y_i = f(x_i; \boldsymbol{\beta}) + \varepsilon_i$$

where f is a known function of known predictor variables x and some unknown parameters $\boldsymbol{\beta}$. For linear regression, we write

$$y_i = \beta_0 + \beta_1 x_{1i} + \ldots + \beta_p x_{pi} + \varepsilon_i$$

That is, the βs are the gradients, or (partial) first derivatives, of the relationship between y and the xs. f is encapsulated by two model statements,

$$E(Y \mid X_1, \ldots, X_p) = \beta_0 + \sum_{j=1}^{p} \beta_j x_{ji}$$

and

$$\mathrm{Var}(Y \mid X_1, \ldots, X_p) = \sigma^2$$

This model is called a linear model because it is linear in the predictors. Formally, a linear model is one for which none of the first derivatives of the mean function with respect to each of the parameters contain any of the parameters. Hence the linearity in the model label refers to the parameters, not to the predictor variables.

Our goal is to find the best estimates of the βs and σ. We have to provide some structure in order to do so. First, we have to choose what is meant by *best* and develop an objective function that reflects the definition. For example, we might say that we would like the estimates of the βs to be those that minimize the sum of the squared residuals. This particular strategy is known as *least-squares* estimation, also referred to as L_2 minimization.

Other objective functions are possible and may be more appropriate. We note in passing that the sum of absolute values of the deviations is an L_1 criterion and the maximum of the absolute deviations is an L_∞ criterion. Least-squares estimation has some pleasant properties that are not immediately obvious, not least among which is that the sums of squares of various portions of the model are additive.

We can construct these estimates using one of numerous methods. To minimize the least-squares objective function, we could use brute force (numerical minimization, using for example the Gauss–Newton algorithm) or, more elegantly, we could differentiate the sum of squared residuals with regard to the βs, set these derivatives to 0, and solve the resulting system of simultaneous equations, called the *normal* equations. Neither of these approaches is taken by R because neither is necessarily numerically stable or efficient; by default R uses QR-decomposition (Chambers and Hastie, 1992, Section 4.4.2).

It turns out that the least-squares parameter estimates are unbiased (see, among many others, Weisberg, 2005), although we did not specify that they had to be. Furthermore, the Gauss–Markov Theorem (see, among many others, Casella and Berger, 1990) establishes that among all linear unbiased estimates the least-squares estimates have the lowest possible variance; that is, they are the most efficient. So, the least-squares estimates are referred to as BLUEs, that is, Best Linear Unbiased Estimates.

However, if the population from which the residuals are sampled is normally distributed, then other useful facts hold. First, the least-squares estimates are also normally distributed, which greatly simplifies the problem of creating and interpreting interval estimates of the parameters. Second, the least-squares estimates are now the minimum-variance estimates among all possible unbiased estimators, not just the linear ones. Such estimates are called minimum-variance unbiased estimators (MVUEs). These latter points provide us motivation to check the assumption of the normality of the residuals; if the regression assumptions are satisfied, then our model has better statistical properties than if they are not satisfied. We may be able to obtain unbiased estimates that have as low a variance as least-squares estimates using some other algorithm, but not unbiased estimates that have lower variance.

Forming interval estimates and performing hypothesis tests on the population parameters both require an estimate of the conditional variance, σ^2. This term can be estimated using the variance of the residuals. However, the model assumes that the variance is constant. There are many ways that this assumption could fail to be true, but in practice we really only worry about one way: if the variance seems to change systematically with any of the available predictor variables. If the variance changes systematically with a variable, then we have to ask whether or not the interval estimates or the outcomes of tests that we compute might change depending on X. That would be an unsatisfactory outcome. Hence we check the assumption of constant variance and take action if the assumption seems unreasonable. For example, we may elect to use generalized linear modeling, or we may transform the response variable, or use weighted least squares, or even try to model the variance explicitly, if we have enough data and sufficient motivation.

A further concern in estimating the residual variance is that we need to be sure that the data are valued appropriately; that is, the observations tell us as much about the system as we hope they do. A key way in which this assumption can fail to be true is if the population from which the residuals are sampled has correlation structure, a situation that is usually informally referred to as having correlated residuals. Although that seems like it might be a simple case to check, of course there are infinite ways that the residuals can be correlated, so again we only worry about a handful of correlation structures, two of which we describe here. First, observations may be similar because they are close together in space or time; for example, trees close together in a forest or measurements of a tree diameter close together in time. This phenomenon is called autocorrelation. Second, observations may be similar because there is a grouping structure and the observations belong to the same group. Examples include pigs in a litter and students in a classroom.

The kind of correlation that we check for is usually suggested by the design of the experiment or the structure of the dataset. In each of the cases, we are sensitive to the possibility of unmodeled influences that are similar for observations that are close or in groups. In practice, a model correction might be made using generalized least squares, or a hierarchical model fit using, for example, mixed-effects models. We discuss generalized least-squares models further in Section 6.1.10 and mixed-effects models in Chapter 7.

Finally, the estimates, tests, and conclusions that we draw all stem from a key assumption that the data actually do represent the population for which we wish to draw inference. If the data as a whole do not represent the population, then the estimates are flawed. In order to provide some protection against this outcome, we check to see whether single values or clusters of values seem to be having an undue effect upon the estimates. Also, we interpret the parameter estimates carefully in the light of our knowledge about the underlying system to be sure that we trust them. No amount of assumption checking will provide any guarantees, however.

6.1.1 Example

We provide a demonstration of the use of regression in R to solve a biological and natural resources problem: the construction of an allometric equation that relates one dimension of an organism with another dimension. We will use this example to motivate and link the following material.

Section 2.4.2 shows the data reading, cleaning, and analysis. We will construct models to predict tree volume in cubic meters using tree diameter measured in cm.

First, we plot the data using the following code. The data are plotted in Figure 6.1. We use the par function with the argument las = 1 to make the tick labels on the y-axis horizontal. Also, note our use of the expression and paste functions in the label for the y-axis. See demo(plotmath) to learn more about this topic.

```
> par(las = 1)
> plot(vol.m3 ~ dbh.cm,
+       data = sweetgum,
```

```
+        xlab = "Diameter (cm)",
+        ylab = expression(paste("Volume (", m^3, ")")))
```

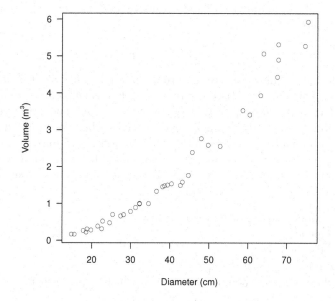

Fig. 6.1: Scatterplot of volume against diameter for sweetgum data.

It is tempting at this point to immediately start thinking of ways that the linear regression could fail and to try to identify the assumptions that are likely to be contradicted. Some analysts may even begin with favored transformations of the response variable. We try to avoid transformation if at all possible. First, it is unusual, although not impossible, for transformation to preserve the meaning of the model. That is, we mind the fact that the model can play an important role as a mathematical representation of the system that is being modeled. Transformation may result in a model with better statistical properties but lacking the meaning inherent in the original model. Second, transformation may not be necessary, depending on the model being fitted and the dataset. Fitting models is inexpensive, and much can be learned about the data and model from fitting the basic models to the raw data and examining them. Also, it may well be that the base model and the raw data combine in an acceptable way. So, we avoid tempting pre-emptive measures. However, fitting models should not be used as a substitute for thinking about the problem.

6.1.2 Thinking about the Problem

We need to identify the scope of applications that the model will be used for. In some settings, the scope of applications will be known, at least approximately, before the data are collected. Often the scope will change, or not be fully formulated before modeling begins, and compromises will be called for.

Here, we wish to be able to predict an unobserved volume conditional on a known diameter,

$$v_i = \beta_0 + \beta_1 \times d_i + \varepsilon_i \qquad (6.1)$$

We would like to make an interval prediction and would like this interval prediction to have good statistical properties. For example, we would like our interval to be wide enough that it covers the true value with specified confidence but be no wider than it really needs to be. We would like our estimators to be unbiased. We would like our estimators to make as efficient use of the data as possible, meaning that they should have minimum variance among the class of unbiased estimators. To support our stated preferences, we will make the following assumptions:

1. the linear model captures the relationship,
2. ε_i are independent,
3. ε_i have constant variance,
4. ε_i are normally distributed, and
5. the sample represents the population from which it was drawn.

The scope of applications of the model is important to know because it affects what compromises we are willing to make in the modeling process, which in turn affects how important the different model assumptions are. As noted above, two fairly distinct applications for models are the prediction of future values and the estimation of population or process parameters. Any application that requires knowledge of the distribution of the estimator, such as hypothesis testing and interval estimation, will demand that the assumptions about constant variance and normal residuals be satisfied to some degree unless the analyst wishes to fall back on large-sample theory. Applications that merely require point estimates will not be so stringent.

6.1.3 Fitting the Model

The lm function is used to fit least-squares regression.

```
> sweetgum.lm.d <- lm(vol.m3 ~ dbh.cm, data=sweetgum)
```

This code fits the model $v_i = \beta_0 + \beta_1 \times d_i + \varepsilon_i$. Note that we did not specify the intercept term; R includes it by default. To remove the intercept from the model, we would instead write `vol.m3 ~ dbh.cm - 1`.

6.1.4 Assumptions and Diagnostics

We noted above that, in order for our estimators to have good statistical properties, we need several assumptions to be true. Each desirable property can be connected to one or more assumptions. Not all properties are necessary to use the model; therefore, some assumptions will be less important than others, depending upon the application.

There is no way to know whether these assumptions are true, but it is possible to determine whether or not the assumptions are reasonable by using graphical diagnostics. Checking assumptions in this way distinguishes a statistical analysis from the mathematical operation of just minimizing an objective function.

The default graphical display, which is created by the `plot` function when an object of class lm is passed as its first argument, produces four plots, and two others can be obtained by using the `which` argument (Figure 6.2). Opinions vary among statisticians as to the utility and necessity of the following graphics. We find them useful. We obtain the diagnostics using the following code. Note the use of `par` to set up a 2×2 matrix of plots with appropriately sized margins.

```
> par(mfrow = c(2, 2), mar=c(4, 4, 3, 1), las = 1)
> plot(sweetgum.lm.d)
```

The default graphical representation of the model provides plots of the model residuals, scaled and manipulated in various ways as described below. Smooth lines are superimposed to aid interpretation; however, in cases where the sample size is small, these smooth lines can overemphasize departures from acceptable patterns, so they should not be overinterpreted. As a remedy, R will allow the `plot` function to pass those arguments that should be passed to the functions that it calls. Hence, a less wiggly set of smooth lines can be obtained by including, say, `span = 2` in the call to `plot`. Also, keep in mind that unstandardized residuals will have non-constant variances by design.

We interpret these diagnostic graphics as follows.

- The top-left panel shows a plot of the residuals against the fitted values, with a smooth curve superimposed. Here we are looking for evidence of curvature and outliers. Our plot shows curvature, which we are concerned about, but no points that are particularly unlike the others. Curvature in this plot is of concern because it suggests possibly substantial local bias in the model. Also, curvature contradicts the first assumption above. These

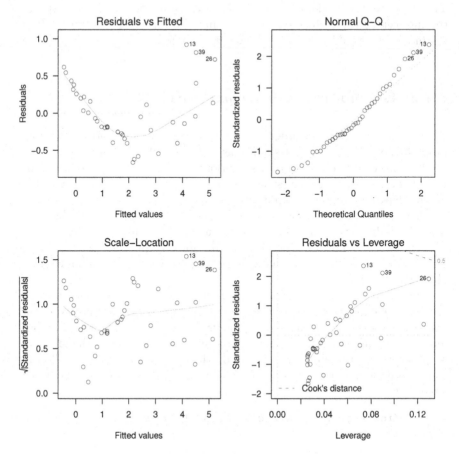

Fig. 6.2: Diagnostic plots for the regression of volume against diameter.

residuals are not standardized, so any non-constant variation in this plot does not concern us.

- The top-right panel shows a quantile-quantile (q-q) plot of the standardized residuals against the normal distribution. Here the ideal plot is a straight line, although modest departures from straightness are often acceptable (due to large-sample theory). Departures from a straight line in this plot may indicate non-normality of the residuals *or* non-constant variance, *or* both. These conditions are imposed by the third and fourth assumptions above. Here, all points are in a reasonably straight line.
- The bottom-left panel shows the square root of the absolute residuals against the fitted values, along with a smooth line. Departures from a horizontal line signify heteroskedasticity, contradicting the third assumption. Here we have modest evidence of changing variance, but it is not consis-

tent. Interpretation of this graph will be complicated by variability in the mean of the residuals, which we clearly observe in the top-left panel.

- The bottom-right panel shows a plot of the leverage of the observations against the standardized residuals. These are the two components of Cook's distance, which is a statistic that reports the overall effect on the parameter estimates of each of the observations (Cook, 1977). So far as observations are concerned, being a large residual or a high leverage point alone is no guarantee of having a substantial impact on the parameter estimates. A reasonably well accepted rule of thumb is that Cook's distances greater than 1 should attract our attention. Contours of these distances[1] at 0.5 and 1.0 are added to the graph by default to assist such an interpretation. This plot provides us with some information about the fifth assumption above. Here we see no evidence to suggest that more than one population has been sampled or that the population that has been sampled is unduly heterogeneous.

Interpretation of diagnostics is as much an art as it is a science. On average, our experience is that more experienced data analysts tend to be less concerned than less experienced analysts about deviation from the strict expectations. Whether this greater relaxation reflects experience or fatigue remains a point of conjecture. However, training one's eye is instructive and easy: generate a clutch of diagnostic plots, for example q-q plots, on data generated randomly under the null hypothesis.[2]

R does not automatically provide diagnostics that are relevant to testing the assumption of independence. If we suspect that residual dependence is a possibility, then we would fit a model that accommodates it and examine estimates of its magnitude. For example, we might check for clustering by fitting a mixed-effects model and estimating the intra-class correlation (see p. 241) and check autocorrelation by fitting a generalized least-squares model with an autocorrelation function using the `gls` function of the nlme package (see Section 6.1.10).

Overall, the panels show worrying curvature. We will now quash our evil thoughts about transformations. In fact, it is natural to think of the following allometric function in the context of predicting volume from diameter:

$$y_i = \beta_0 x_i^{\beta_1} \times \varepsilon_i \tag{6.2}$$

It is easy to fit something like this model, although with additive instead of multiplicative errors, as a non-linear least-squares model (see Section 6.2.9). For the moment, we take log transformations of the response and the predictor variables and use those as the predictor and response variables for a new linear model.

---

[1] These contours should probably be referred to as *isoCooks*.

[2] For example, `par(mfrow=c(4,4)); for (i in 1:16) qqnorm(rnorm(20))`.

```
> sweetgum$log.vol.m3 <- log(sweetgum$vol.m3)
> sweetgum$log.dbh.cm <- log(sweetgum$dbh.cm)
> sweetgum.lm.ld <- lm(log.vol.m3 ~ log.dbh.cm,
+                         data = sweetgum)
```

Again, we examine the model diagnostics (Figure 6.3).

```
> par(mfrow = c(2, 2), mar=c(4, 4, 3, 1), las=1)
> plot(sweetgum.lm.ld)
```

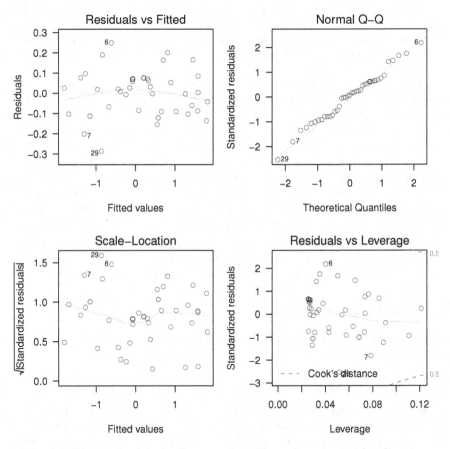

Fig. 6.3: Diagnostic plots for the regression of log volume against log diameter.

Overall, these diagnostics are much more acceptable than those presented in Figure 6.2. There is some evidence of lack of fit in the top-left panel, but it is probably not worth worrying about for the moment. It might be worth experimenting with a few other terms in the model to see if the lack of fit can be ironed out, but to do so here is beyond the scope of this chapter.

6.1.5 Examining the Model

Having satisfied ourselves that the model is at least reasonable, by using the regression diagnostics, we can examine parameter estimates and summary statistics. The summary function reports a number of useful statistics. We dissect it as follows.

```
> summary(sweetgum.lm.1d)
```

First, the fitted model call is identified. This identification may seem redundant, but it is useful when the model is more complex and has been specified using some of the convenient shortcuts that are available.

```
Call:
lm(formula = log.vol.m3 ~ log.dbh.cm, data = sweetgum)
```

The distribution of the residuals is then summarized by the minimum, maximum, median, and two quartiles. Ideally the distribution will be symmetric.

```
Residuals:
     Min        1Q     Median        3Q        Max
-0.286306 -0.089245   0.008477   0.071497   0.249096
```

A regression table is presented, including the parameter estimates, their estimated standard errors, the t-statistic for the two-sided null hypothesis that the population or process parameter is actually 0, and the p-value appropriate to that same test.

```
Coefficients:
             Estimate Std. Error t value Pr(>|t|)
(Intercept) -7.80452    0.14542   -53.67   <2e-16 ***
log.dbh.cm   2.22675    0.04023    55.35   <2e-16 ***
---
Signif. codes:  0 '***' 0.001 '**' 0.01 '*' 0.05 '.' 0.1 ' ' 1
```

Finally, some summary statistics of the model fit are presented: the estimated standard deviation of the errors, the multiple and adjusted R-squared values, and the F-statistics and associated p-value for the null hypothesis that the model is no better than the sample mean of the response variable.

```
Residual standard error: 0.1163 on 37 degrees of freedom
Multiple R-squared: 0.9881,      Adjusted R-squared: 0.9877
F-statistic:  3063 on 1 and 37 DF,  p-value: < 2.2e-16
```

An informal way to assess the usefulness of the model is to compare the marginal and conditional variabilities of the response variable — e.g., compare the standard deviation of the response variable with the standard deviation of the residuals. Here there is a big difference between the marginal variability, which is 1.05 (units are $\log m^3$), and the conditional variability, which is 0.12 (units are $\log m^3$). The model appears to be a useful one. A formal comparison would require the use of cross-validation or a similar strategy (see, e.g., Hastie et al., 2009).

We can obtain interval estimates of the model parameters in several ways. For example, we can compute estimates directly from a portion of the output of the summary statement. Here we capture the output as an object, convert it to a data frame, and then manipulate it to obtain an interval estimate for the slope.

```
> reg.table <- as.data.frame(coef(summary(sweetgum.lm.1d)))
> reg.table$Estimate[2] +
+    reg.table$`Std. Error`[2] *
+    qt(c(0.025, 0.975), summary(sweetgum.lm.1d)$df[2])

[1] 2.145230 2.308270
```

The same quantities can be more easily obtained using the `confint` function:

```
> confint(sweetgum.lm.1d)

               2.5 %     97.5 %
(Intercept) -8.099178 -7.509869
log.dbh.cm   2.145230  2.308270
```

In order to interpret the interval estimates appropriately, it is essential that we check the assumptions that underpin them. Specifically, the interval estimates are made conditional on the assumption that the sampling distribution of the parameter estimate is normal with mean equal to the population parameter and constant variance. For this to be true, we need the assumptions 1–5 listed on p. 180 to be true or that the sample size be sufficiently large that the Central Limit Theorem can be invoked in place of assumption 4 . These assumptions are assessed using the graphical diagnostics presented in Figure 6.2 as well as our knowledge about the system or process that the data represent and our knowledge of the protocols by which the data were collected.

More general manipulation of the parameter estimates is also possible. Although it seems cumbersome, the `estimable` function of the gmodels package provides great flexibility in that it constructs point and interval estimates for arbitrary linear combinations of the parameter estimates for a wide range of different model objects (Warnes, 2010). The function requires a specification of the linear combinations of interest, but the specification can be in one of a

number of formats. Here we opt for a matrix with named rows and truncate
the output. See the `estimable` help file for more information.

```
> library(gmodels)
> estimable(sweetgum.lm.ld,
+           rbind(Intercept = c(1,0),
+               Slope     = c(0,1)),
+          conf.int = 0.95)[,c(1,2,4,6,7)]

          Estimate Std. Error DF  Lower.CI  Upper.CI
Intercept -7.804523 0.14542291 37 -8.099178 -7.509869
Slope      2.226750 0.04023317 37  2.145230  2.308270
```

We now diverge from the example to discuss elements of R that are suggested by the preceding code. Readers that are uninterested in the internals
of R should skip to the next section.

We can identify some of the object classes that have a corresponding
`estimable` method via the **methods** function.

```
> methods(estimable)

[1] estimable.default* estimable.mer*     estimable.mlm*

    Non-visible functions are asterisked
```

However, the list is not definitive because the default function may handle
a wide range of object types. What do we do? The next step would be to
examine the source code of the default method by printing it,

```
> estimable.default

Error in eval(expr, envir, enclos) : object 'estimable.default'
not found
```

but the code seems to be hidden. Therefore we call on the **getAnywhere**
function with success.

```
> getAnywhere(estimable.default)

A single object matching 'estimable.default' was found
It was found in the following places
  registered S3 method for estimable from namespace gmodels
  namespace:gmodels
with value

function (obj, cm, beta0, conf.int = NULL, show.beta0,
    joint.test = FALSE, ...)
```

And, scrolling down through the code, we find the list of classes that the
function is used for.

```
"obj must be of class 'lm', 'glm', 'aov', 'lme', 'gee', 'geese'
    or 'nlme'"
```

We have included this exercise because it is an instructive demonstration of the kind of digging that is required to be able to use R effectively. We can learn a great deal about the environment from scrutiny. Questions that seem insoluble can often be easily resolved by examining the object and asking straightforward questions about it: What are its classes, what are its dimensions, and what are its contents?

6.1.6 Using the Model

Our goal in constructing this model was to provide a means to obtain a prediction of volume for a tree for which we know only the diameter.

We note that the effect of fitting an unbiased model to log-transformed data and back-transforming to the natural scale results in a model that forms biased predictions (see Jensen, 1906). Flewelling and Pienaar (1981) provide some discussion and candidate corrections. The parameters that are used by the corrections have to be estimated from the model in any case, and this estimation will add uncertainty to the predictions. Here an estimate of the multiplicative correction is

$$\exp\left(\frac{\hat{\sigma}^2}{2}\right) \tag{6.3}$$

```
> exp(summary(sweetgum.lm.ld)$sigma^2 / 2)
```

```
[1] 1.006784
```

using the second candidate of Flewelling and Pienaar (1981). The reader may wish to estimate the confidence interval for the correction, bearing in mind that R provides the inverse cdf for the χ^2 distribution as qchisq and that $(n-p)s^2/\sigma^2 \sim \chi^2_{n-p}$. We also note the high importance of the assumption of constant variance for the sensible application of this bias correction.

We can write a function to ease the use of the model for making corrected predictions. This function will accept one or more diameters, in cm, and predict the corresponding volume or volumes, approximately corrected for bias.

```
> sweetgum.vol.hat <- function(dbh.cm,
+                              ht.dbh.lm = sweetgum.lm.ld,
+                              correct = TRUE) {
+     sweetgum.hat <- data.frame(dbh.cm = dbh.cm)
+     sweetgum.hat$log.dbh.cm <- log(sweetgum.hat$dbh.cm)
+     correction.2 <- ifelse(correct,
```

```
+                          exp(summary(ht.dbh.lm)$sigma^2 / 2),
+                          1)
+   sweetgum.hat$vol.m3 <-
+     exp(predict(ht.dbh.lm, newdata = sweetgum.hat)) *
+       correction.2
+   return(sweetgum.hat$vol.m3)
+ }
```

We can then use the function

```
> sweetgum.vol.hat(10, sweetgum.lm.1d)
```

```
[1] 0.06921896
```

as inputs, as well as single values. Such code is called *vectorized* and will execute much more quickly than an equivalent operation within a loop.

```
> sweetgum.vol.hat(c(10,12), sweetgum.lm.1d)
```

```
[1] 0.06921896 0.10388239
```

We can also omit the correction to see how large it is for these trees.

```
> sweetgum.vol.hat(c(10,12), sweetgum.lm.1d, correct = FALSE)
```

```
[1] 0.06875256 0.10318242
```

   We note in passing that we could simplify the use of the function even further by omitting the second argument and identifying our lm object directly in the predict function. R would search the environment that has been created by the function for the object and upon failing to find it would search the parent environment from which the function was called. R would find the object there and either use that object (if no changes are being made) or a copy of it if changes are to be made. This nested search path provides an example of lexical scoping. Lexical scoping permits complicated sets of operations to be expressed and performed very easily. However, lexical scoping contains traps for the unwary; it is easy to perform operations on the wrong object.
   Figure 6.4 presents a view of the data and the model. It replicates Figure 6.1 and adds the fitted line by using the curve function.

```
> par(las = 1)
> plot(vol.m3 ~ dbh.cm,
+       data = sweetgum,
+       xlab = "Diameter (cm)",
+       ylab = expression(paste("Volume (", m^3, ")")))
> curve(sweetgum.vol.hat, from=0, to=100, add=TRUE)
```

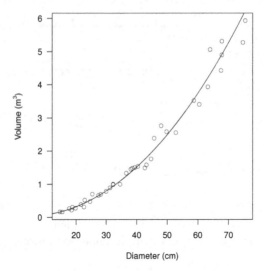

Fig. 6.4: Scatterplot of volume against diameter for sweetgum data with fitted model superimposed.

In order for `curve` to work, we needed to be sure that all arguments for `sweetgum.vol.hat` other than the first were provided default values in the function definition. Default values are provided to the function using the following protocol in the declaration: `argument = default`.

We can use the `predict` function to obtain prediction intervals for values of interest of the predictor variables. For example,

```
> exp(predict(sweetgum.lm.ld,
+             newdata = data.frame(log.dbh.cm = log(c(10, 20)))),
+             interval = "prediction"))

        fit         lwr         upr
1 0.06875256 0.05298514 0.08921207
2 0.32181591 0.25229082 0.41050038
```

These estimates are not corrected for bias, although extending the function above to provide bias-corrected interval estimates is a straightforward exercise.

We now take another brief detour into the workings of R. Again, we direct the disinterested reader to the end of the section.

Model interrogation is enriched if we take advantage of the fact that the model that we have fitted is an object and the summary of that model fit is a different object. The following discussion pertains to S3 classes. Complete coverage of R's facilities for object-oriented programming, including S3 and S4 classes, is beyond the scope of this book.

```
> names(sweetgum.lm.1d)
```

```
  [1] "coefficients"    "residuals"       "effects"
  [4] "rank"            "fitted.values"   "assign"
  [7] "qr"              "df.residual"     "xlevels"
 [10] "call"            "terms"           "model"
```

```
> names(summary(sweetgum.lm.1d))
```

```
  [1] "call"            "terms"           "residuals"
  [4] "coefficients"    "aliased"         "sigma"
  [7] "df"              "r.squared"       "adj.r.squared"
 [10] "fstatistic"      "cov.unscaled"
```

The reader may also wish to try the `str` function on these two objects.

Some high-level functions, called **methods**, exist to enable the reliable extraction of model information, for example, the `residuals` and `fitted` functions are methods that will extract the residuals and the fitted values, respectively.

We can learn what methods are available for our object as follows: first identify the class of the object, and then call the methods function using the class as the argument. Here we print only the first few.

```
> class(sweetgum.lm.1d)
```

```
[1] "lm"
```

```
> head(methods(class = class(sweetgum.lm.1d)))
```

```
[1] "BIC.lm"          "add1.lm"         "addterm.lm"      "alias.lm"
[5] "anova.lm"        "attrassign.lm"
```

We can also extract or otherwise manipulate the attributes of some objects by means of the $ sign:

```
> sweetgum.lm.1d$call
```

```
lm(formula = log.vol.m3 ~ log.dbh.cm, data = sweetgum)
```

```
> summary(sweetgum.lm.1d)$sigma
```

```
[1] 0.1162833
```

Generally, code authors prefer that attributes be extracted using custom-written extractor functions. An example extractor function to extract the estimate of the standard deviation of the residuals might look like this:

```
> sigma <- function(x) {
+         if (class(x) == "lm") {
+            return(summary(x)$sigma)
+         } else {
```

```
+          stop("Object is not a linear model (class 'lm').")
+       }
+ }
```

```
> sigma(sweetgum.lm.1d)
```

```
[1] 0.1162833
```

The advantage of using custom extractors is that the code author can change the object structure and methods as necessary without unduly inconveniencing the code user. An advantage of using such functions is that there is a consistent interface, and options such as standardizing the residuals can be coded and performed as needed. However, extractor functions are not always available.

6.1.7 Testing Effects

Statistical theory offers numerous different approaches for assessing the contribution of a predictor variable to the model, some of which are related. Statistical tests involve, by and large, a comparison of a statistic against a standard. The standard is usually calculated in a way to represent no effect, or more generally the null effect, encapsulated as the null hypothesis. If the statistic and the standard are distinctly different, then the hypothesis is rejected. A key point is that the standard is almost always determined using assumptions.

The assumptions that were important for the qualities of the point estimators of the parameters are also important for the interval estimators and tests.

The summary function for the linear model object provides a coefficient table that includes t-tests of the null hypothesis that the parameter is zero (see p. 185). This specific test may or may not have any useful interpretation in the context of model application. However, the t-test is a way to test whether or not the term *should* be included (for some values of *should*). The table can also be abstracted from the summary object as follows.

```
> summary(sweetgum.lm.1d)$coef
```

```
              Estimate Std. Error   t value      Pr(>|t|)
(Intercept) -7.804523 0.14542291 -53.66777 1.064901e-36
log.dbh.cm   2.226750 0.04023317  55.34614 3.454727e-37
```

Here we see no evidence to suggest that either parameter is redundant in the model.

R also provides the **anova** function, which produces F-tests on the sums of squares that are attributable to each term. If the predictor variables are independent (orthogonal), then the order of the terms does not matter. If they are not orthogonal, then the terms are tested sequentially in descending order; that is, each term is tested assuming that the terms above it are included in the model. The following code produces a standard analysis of variance table.

The following example uses the Upper Flat Creek data, which were introduced in Section 2.4.1.

```
> hd.lm.1 <- lm(I(log(height.m)) ~ dbh.cm * species,
+                data = ufc.tree,
+                subset = height.m > 0)
> anova(hd.lm.1)

Analysis of Variance Table

Response: I(log(height.m))
                 Df Sum Sq Mean Sq  F value     Pr(>F)
dbh.cm            1 40.477  40.477 917.6248  < 2.2e-16 ***
species          11  2.260   0.205   4.6586  7.55e-07 ***
dbh.cm:species   10  1.161   0.116   2.6331  0.003818 **
Residuals       603 26.598   0.044
---
Signif. codes:  0 '***' 0.001 '**' 0.01 '*' 0.05 '.' 0.1 ' ' 1
```

Note the use of the I function in the latter approach. This tells R to interpret the function in the usual way, not in the context of the linear model. The difference becomes more obvious when we think about the double usage that R is placing upon our familiar operators. For example, * usually signifies multiplication, but in the linear model it means 'include main effects and interaction'. Likewise, + usually means addition, but in the linear model it is used to separate the additive terms in the model statement. This phenomenon is called *operator overloading*. So, in order to be sure that R will interpret our instructions arithmetically, we wrap them in I.

The anova function can also be used to compare models by using the whole-model test. We fit a simpler model and then compare the simpler and more complex models as follows.

```
> hd.lm.2 <- lm(I(log(height.m)) ~ dbh.cm,
+                data = ufc.tree,
+                subset = height.m > 0)
> anova(hd.lm.1, hd.lm.2)

Analysis of Variance Table

Model 1: I(log(height.m)) ~ dbh.cm * species
Model 2: I(log(height.m)) ~ dbh.cm
  Res.Df    RSS Df Sum of Sq      F    Pr(>F)
1    603 26.599
```

```
2      624 30.020 -21    -3.4219 3.6941 7.594e-08 ***
---
Signif. codes:  0 '***' 0.001 '**' 0.01 '*' 0.05 '.' 0.1 ' ' 1
```

R uses an algorithm to control the order in which terms enter a model. The algorithm automatically respects parameter hierarchy in that all interactions enter the model after the main effects, and so on. However, sometimes we wish to change the order in which terms enter the model. This may be because we are interested in testing certain terms while other terms are included in the model. For example, using the **npk** dataset provided by MASS (Venables and Ripley, 2002), imagine that N and P are design variables and we wish to test the effect of K in a model that already includes the N:P interaction.

```
> require(MASS)
> data(npk)
> anova(lm(yield ~ block + N * P + K, npk))

Analysis of Variance Table

Response: yield
          Df Sum Sq Mean Sq F value   Pr(>F)
block      5 343.30  68.659  4.3911 0.012954 *
N          1 189.28 189.282 12.1055 0.003684 **
P          1   8.40   8.402  0.5373 0.475637
K          1  95.20  95.202  6.0886 0.027114 *
N:P        1  21.28  21.282  1.3611 0.262841
Residuals 14 218.90  15.636
---
Signif. codes:  0 '***' 0.001 '**' 0.01 '*' 0.05 '.' 0.1 ' ' 1
```

The preceding code fails to provide tests of the terms of interest in the desired order because of R's effect-order algorithm. To force the order in which the effects enter the model, we must use the **terms** function.

```
> anova(lm(terms(yield ~ block + N * P + K,
+                 keep.order = TRUE), npk))

Analysis of Variance Table

Response: yield
          Df Sum Sq Mean Sq F value   Pr(>F)
block      5 343.30  68.659  4.3911 0.012954 *
N          1 189.28 189.282 12.1055 0.003684 **
P          1   8.40   8.402  0.5373 0.475637
N:P        1  21.28  21.282  1.3611 0.262841
K          1  95.20  95.202  6.0886 0.027114 *
Residuals 14 218.90  15.636
```

```
---
Signif. codes:  0 '***' 0.001 '**' 0.01 '*' 0.05 '.' 0.1 ' ' 1
```

Happily, we learn that the order of the terms makes no difference to our inference in this case. A balanced dataset is a wonderful thing!

6.1.8 Transformations

Sometimes the modeler has biological or statistical reasons to change the nature of the data that are being modeled. Such transformations can be done either by creating a new variable in the data frame with the appropriate function or by making the transformation inside the call to the model.

```
> ufc.tree$log.height.m <- log(ufc.tree$height.m)
> hd.lm.4a <- lm(log.height.m ~ dbh.cm * species,
+               data = ufc.tree,
+               subset = height.m > 0)
> hd.lm.4b <- lm(I(log(height.m)) ~ dbh.cm * species,
+               data = ufc.tree,
+               subset=height.m > 0)
```

Transformation can be very useful, but it is not without cost. We believe that if possible it is best to work in the units that will make the most sense to the person who uses the model. Of course, it is possible to back-transform arbitrarily, and correct for bias, but why do so if it is unnecessary? There are certainly circumstances where transformation is called for, but often a more appropriate and satisfying strategy is available, whether that be fitting a generalized linear model or an additive model or performing a small Monte Carlo experiment on the residuals and placing trust in the Central Limit Theorem.

6.1.9 Weights

We have been analyzing the UFC data assuming that each observation is equally informative about the population. However, although the sample trees may be considered a random sample, they are not selected with equal probability, and furthermore they are clustered in location because they were selected using a variable-radius plot sample. How much difference to the parameter estimates does this sample design make? Here we will assess the effect of the sample weights. We will examine tools that more formally illuminate the effect of clustering upon our parameter estimates in the next chapter.

Recall that, in a variable-radius plot, the probability of selection of a tree is proportional to its basal area. We can fit a model that accommodates this effect as follows.

```
> hd.lm.5 <- lm(height.m ~ dbh.cm * species,
+                weights = dbh.cm^-2,
+                data = ufc.tree)
```

Now we may wish to know what effect this change of weights has on our parameter estimates. To pull out estimates of each slope and intercept, we can use the **estimable** function provided in the gmodels package, but for models that have only one categorical predictor variable, there is a hack, a simpler way to obtain them. We refit the models with the intercept and, if necessary, the slope explicitly removed from the model statement. That is,

```
> unweighted <-
+   coef(lm(height.m ~ dbh.cm * species - 1 - dbh.cm,
+           data = ufc.tree))
> weighted <-
+   coef(lm(height.m ~ dbh.cm * species - 1 - dbh.cm,
+           weights = dbh.cm^-2, data = ufc.tree))
```

Note that altering the model in this way (that is, fitting the model with the intercept omitted) distorts the meaning of much of the summary output, so we advocate its use only for expediting the extraction of specific parameter estimates.

We can now plot these estimates in various more or less informative ways. The following code constructs Figure 6.5, which provides a compact graphical summary of the effects of weighting. We use the **text** function to place symbols and the **abline** function to provide an $x = y$ line for the purposes of comparison.

```
> intercepts <- 1:10; slopes <- 11:20
> par(mfrow = c(1,2), las = 1, mar = c(4,4,3,1))
> plot(unweighted[intercepts], weighted[intercepts],
+       main = "Intercepts", type = "n",
+       xlab = "Unweighted", ylab = "Weighted")
> abline(0, 1, col="darkgrey")
> text(unweighted[intercepts], weighted[intercepts],
+       levels(ufc.tree$species))
> plot(unweighted[slopes], weighted[slopes],
+       main = "Slopes", type = "n",
+       xlab = "Unweighted", ylab = "Weighted")
> abline(0, 1, col = "darkgrey")
> text(unweighted[slopes], weighted[slopes],
+       levels(ufc.tree$species))
```

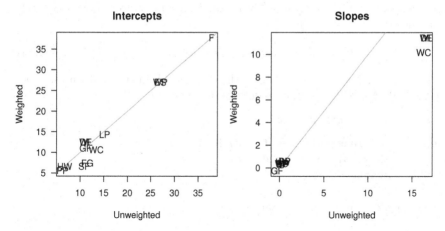

Fig. 6.5: Summary of effect of sample weights upon species-specific height–diameter parameter estimates for UFC data.

The effect of weighting upon the parameter estimates appears to be pretty substantial. This effect can be interpreted as telling us that the smaller and the larger trees require different models because the parameter estimates change when the small trees are weighted more heavily.

6.1.10 Generalized Least-Squares Models

Another solution to the problems of non-constant variance and auto-correlated residuals is to explicitly include those elements in the model. If these elements are included as components of the model, then the assumptions are more likely to be satisfied, the parameter estimators will have better statistical properties, and the model may be more informative.

One convenient way to fit linear models that include these components is to use the `gls` function of the nlme package (Pinheiro and Bates, 2000). The model expression and manipulation is largely identical to that of `lm`, but other arguments are possible. For example, to add a fitted variance model to the sweetgum volume function, we would use the code

```
> library(nlme)
> sweetgum.gls.ld <- gls(log.vol.m3 ~ log.dbh.cm,
+                        weights = varPower(form = ~ dbh.cm),
+                        data = sweetgum)
```

The model that this code has fitted is the same as above, except that we no longer assume that the variance of the residuals is constant. Instead, we assume that the variance of the residuals is some power of the diameter. Specifically, we assume that

$$\text{Var}(\varepsilon_i) = \sigma^2 |d_i|^{2\delta} \tag{6.4}$$

where σ^2 and δ are estimated from the data (see Pinheiro and Bates, 2000, p. 210). We can examine the model in the usual way.

```
> summary(sweetgum.gls.1d)
```

```
Generalized least squares fit by REML
  Model: log.vol.m3 ~ log.dbh.cm
  Data: sweetgum
        AIC       BIC    logLik
  -41.61428 -35.17060 24.80714


Variance function:
 Structure: Power of variance covariate
 Formula: ~dbh.cm
 Parameter estimates:
     power
-0.2801518


Coefficients:
                Value  Std.Error   t-value p-value
(Intercept) -7.760255 0.15079627 -51.46185       0
log.dbh.cm   2.214602 0.04045249  54.74575       0


 Correlation:
           (Intr)
log.dbh.cm -0.993


Standardized residuals:
       Min          Q1        Med          Q3         Max
-2.24372393 -0.80677668 0.09360657  0.56845164  1.93350960


Residual standard error: 0.3124869
Degrees of freedom: 39 total; 37 residual
```

The variance function is now taken into account for all reported summaries of the model. Standardized residuals will be standardized conditional on this model, for example. They are obtained by including the type = "pearson" argument to the call to residuals.

We can even obtain an approximate test of the improvement of the quality of the model by using the anova function, so long as we include the gls-fitted object as the first argument.

```
> anova(sweetgum.gls.1d, sweetgum.lm.1d)
```

```
              Model df      AIC        BIC   logLik   Test
sweetgum.gls.ld    1  4 -41.61428 -35.17060 24.80714
sweetgum.lm.ld     2  3 -42.44004 -37.60728 24.22002 1 vs 2
              L.Ratio p-value
sweetgum.gls.ld
sweetgum.lm.ld  1.174237  0.2785
```

The *p*-value above suggests poor support in the data for the inclusion of a variance function that depends on the predictor variable.

The `gls` class has a number of useful methods. Use

```
> methods(class = "gls")
```

to discover them. Pinheiro and Bates (2000) is essential reading.

6.2 Non-linear Regression

One of the unsatisfying characteristics of linear models is that we are constrained in how the parameters and coefficients can interact. We have had to ensure that the normal equations, which are the first derivatives of the objective function with respect to the coefficients, were independent of the coefficients. This constraint has precluded us from using a wealth of biologically realistic model forms. There have also been advantages: we have been guaranteed that any inference has been exact, conditional on the assumptions and the model, and that fitting has required one step only and no starting values. However, it is arguably better to get an approximate answer to a meaningful question than to get an exact answer to an approximation to a meaningful question.[3]

With *non-linear* model fitting, we can deploy classes of models that do not have such a restriction. This broader stance allows us to choose model forms that we feel genuinely represent the underlying processes in some way, which in turn allows us to think about the fitting process in more than purely empirical ways.

As before, we do not present the theory here, as that has been covered more effectively elsewhere. Recommended reading includes Ratkowsky (1983), Gallant (1987), Bates and Watts (1988), Schabenberger and Pierce (2002), and Seber and Wild (2003).

Examples of non-linear models that might be relevant in forestry and natural resources include:

- change-point models
- plateau models
- allometric models

---

[3] This quotation is a first-order approximation to Tukey (1962).

- growth models

Schabenberger and Pierce (2002) charge that researchers tend to avoid non-linear models because of discomfort with the messiness of implementation. The messiness is unavoidable; however, non-linear models are parsimonious, incorporate limits, and can be parameterized so that parameters will have a direct interpretation.

We consider non-linear models with one continuous predictor. We write

$$y_i = f(\mathbf{x_i}, \beta) + \varepsilon_i \qquad (6.5)$$

where β is a $p+1$-length vector of parameters; e.g., $\beta = (\beta_0, \beta_1, \ldots, \beta_p)'$.

This model statement is sufficient to find least-squares estimates: we simply need to find those values of β that minimize the sum of squared deviations between the observed y_i and the fitted y_i. Furthermore, the least-squares estimates of the parameters are asymptotically normally distributed if the following assumptions hold:

1. $\mathrm{Var}(\varepsilon_i) = \sigma^2$, a constant.
2. ε_i are independent.
3. ε_i are normally distributed.

As in the linear model case, each of the assumptions play different roles and are of varying importance depending on the purpose to which the model will be put. Note that the assumption of zero mean for the errors depends on whether or not an additive constant is included in f.

6.2.1 Example

As in the previous section, we will motivate non-linear modeling via an example. Our data are Norway spruce measurements drawn from von Guttenberg (1915), kindly provided to us by Professor Boris Zeide. Our goal is, in the first instance, to construct a model that predicts tree diameter as a function of age. The data import and cleaning are documented in Section 2.4.4.

6.2.2 Thinking about the Problem

We start with a plot of the trajectories of diameter growth. The following code uses Hadley Wickham's `qplot` function from the ggplot2 package in R, documented in Wickham (2009), to produce Figure 6.6.

```
> library(ggplot2)
```

```
> qplot(x = age.bh, y = dbh.cm, group = tree,
+        linetype = site, facets = ~ location,
+        xlab = "Age (y)", ylab = "Dbh (cm)",
+        geom = "line",
+        data = gutten) +
+   scale_x_continuous(breaks = 40 * (0:3)) +
+   scale_y_continuous(breaks = 10 * (0:5))
```

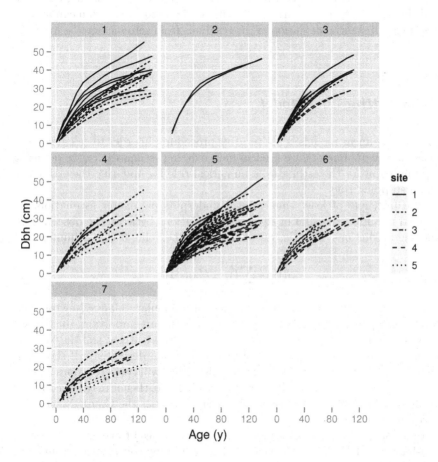

Fig. 6.6: Diameter–age trajectories for von Guttenberg's Norway spruce data. The panels represent locations and the line types represent sites.

The reader may prefer to substitute color = site for linetype = site in the code.

After inspecting Figure 6.6, we shall use a simple non-linear model that passes through the origin and provides an asymptote. Passing through the origin is reasonable as a starting point because, by definition, dbh is non-zero only after age at breast height is non-zero. That may prove to be an ugly constraint, however. A candidate model is then

$$y_i = \phi_1 \times \left[1 - \exp\left(-\frac{x_i \log 2}{\phi_2} \right) \right] + \varepsilon_i \qquad (6.6)$$

where ϕ_1 is the fixed, unknown asymptote and ϕ_2 is the fixed, unknown time until the tree reaches half its predicted maximum size (see Pinheiro and Bates, 2000, who use a different parameterization). Note that the model omits an additive constant, so we can't expect the residuals to have zero mean and we may take substantial deviation of the residuals from zero mean to be evidence of lack of fit.

6.2.3 Fitting the Model

Fitting non-linear models is an iterative procedure; in order to calculate estimates of the parameters, we need to have existing estimates. So, we require a starting point for the estimates, a numerical rule for updating the estimates, and a stopping rule.

The principle of least squares still underpins parameter estimation: we wish to minimize the sum of squares of deviations between the observations and the predictions. That is, we want to find the value $\hat{\theta}$ that minimizes

$$S(\theta) = \sum_{i=1}^{n} (y_i - f(\mathbf{x}_i, \beta))^2$$

Two parameter-estimation strategies are popular: Gauss–Newton and Newton–Raphson. The differences between the two are important: the Gauss–Newton algorithm requires the calculation of first derivatives, and the Newton–Raphson requires the calculation of first and second derivatives. If the residuals are small, then the Gauss–Newton algorithm will converge more rapidly than Newton–Raphson, but if the residuals are large, then it may not converge at all. R's non-linear model fitting function, nls, uses the Gauss–Newton algorithm by default.

In order to fit a non-linear model, we generally need to construct our own non-linear model function. We can do this simply by constructing a function that produces the appropriate predictions given the inputs. However, fitting the model turns out to be more efficient if we can also pass the first and maybe second derivatives to the function as well. Writing a function to pass the predictions and the derivatives is simplified by using the deriv function. NB: R provides some simple prepackaged models, about which we will learn more later.

Our use of `deriv()` requires three things: a statement of the function, a vector of parameter names for which derivatives are required, and the template of the function call.

```
> dbh.growth <-
+    deriv(~ asymptote * (1 - exp(-log(2)/scale * x)),
+          c("asymptote","scale"),
+          function(x, asymptote, scale){},
+          hessian = TRUE)
```

The Hessian is the matrix of estimated second partial derivatives. We asked for the Hessian to be returned as well because it will be used later in diagnostic code that determines the curvature of the model.

Having written the non-linear model as a function, we should try it out. We select a handy tree:

```
> handy.tree <- subset(gutten, tree.ID == "1.1")
```

In a departure from our earlier chapters, we also need to guess the starting estimates for the parameters. Here we will use the highest value as an estimate of the asymptote and guess that the tree reaches about half its maximum diameter in about 30 years.

```
> max(handy.tree$dbh.cm, na.rm=TRUE)
```

```
[1] 29
```

Then the model is fit using `nls`, as per the following code. Note the inclusion of the argument `na.action = na.exclude` to tell R how to handle missing values: we are asking R to take the missing values out of the data prior to fitting but also to pad its observation-level statistics, such as fitted values and residuals, so that they are of the same length as the original data. We will need this facility in the construction of Figure 6.7.

```
> handy.nls <-
+    nls(dbh.cm ~ dbh.growth(age.bh, asymptote, scale),
+        start = list(asymptote = 29, scale = 10),
+        na.action = na.exclude,
+        data = handy.tree)
```

6.2.4 Assumptions and Diagnostics

We now need to assess the model assumptions in the light of the fit of the model to the data. Some background will help us to understand the importance of these assumptions.

Geometry provides a convenient way to think about least-squares linear regression. Briefly, the response variable is an n-dimensional vector in n-space. The predictor variables are used to form a p-dimensional surface, specifically a hyperplane, in the n-space. Then, the model predictions are the p-dimensional vector within the p-dimensional hyperplane that is closest to the n-dimensional vector in n-space. That p-dimensional vector is found by projecting the n-dimensional vector into the p-dimensional hyperplane. The parameter estimates are then the coordinates of that p-dimensional vector.

In non-linear least squares, the least-squares geometric principles are assumed to hold locally, but they are known to hold only approximately. The quality of the information about the model fit depends on how close to linear the model is, for the data, in the region of the optimum.

First, there is no guarantee that the surface that is formed by the predictor variables is flat, or the multi-dimensional equivalent of flat. This lack of flatness is because the values of the parameters affect the shape of the model; that is, the normal equations are not independent of the parameters. Therefore projection will not find the nearest point; the location upon the hyperplane of the nearest point to the n-vector must be found iteratively. The assumption of local flatness is referred to as the *planar assumption* (Seber and Wild, 2003, p. 134) and is estimated using the *intrinsic curvature*.

Second, there is no guarantee that the surface thus formed has well-behaved coordinates; that is, coordinates that map to straight, parallel, equidistant lines. The scale may change along one direction or another, the coordinates might be curved, and the inter-coordinate distance may vary. The assumption of locally well-behaved coordinates is called the *uniform coordinate assumption* (Seber and Wild, 2003, p. 135), and is estimated using the *parameter-effects curvature*.

We will refer to these assumptions jointly as the local-linearity assumptions.

The MASS package in R (Venables and Ripley, 2002) provides code that can be used to compute the intrinsic and the parameter-effects curvatures for a given model, so long as the model function can provide the Hessian. The code, which we invoke after loading the MASS package, is

```
> rms.curv(handy.nls)
```

```
Parameter effects: c^theta x sqrt(F) = 0.096
        Intrinsic: c^iota  x sqrt(F) = 0.0246
```

The value 0.3 has been suggested as a soft cutoff for the point at which the curvature becomes unacceptable for the planar assumption and the uniform coordinate assumption. Both assumptions seem to be reasonable for this model and these data.

As a side note, we also fit the model represented in equation (6.6) using the parameterization suggested in Pinheiro and Bates (2000),

$$y_i = \phi_1 \times [1 - \exp(-\exp(\phi_2)x_i)] + \varepsilon_i \qquad (6.7)$$

An advantage of that parameterization is that the effect of scale is constrained to be positive, so the process of estimation may be more stable. This strategy is also useful for constraining estimates of parameters that have biological constraints. We obtained the following curvature estimates:

```
Parameter effects: c^theta x sqrt(F) = 0.1818
       Intrinsic: c^iota  x sqrt(F) = 0.0246
```

Note that the intrinsic curvature is unchanged. That is, our efforts left the fundamental shape of the surface upon which we were projecting unchanged. However, the parameter-effects curvature has doubled, suggesting that the assumption of uniform coordinates for the Pinheiro and Bates (2000) parameterization does not hold as well *for these data* as the assumption for the previous parameterization. For further reading, see Ratkowsky (1983), Chapter 7 of Bates and Watts (1988), and Section 5.7.1 of Schabenberger and Pierce (2002).

The intrinsic curvature also affects the nature of the residuals in that in the presence of substantial intrinsic curvature the residuals may have nonzero means, non-unit variances, and negative correlation with the response variable (Seber and Wild, 2003, p. 179). Note that

```
> mean(residuals(handy.nls), na.rm=TRUE)
```

```
[1] -0.06936603
```

Therefore residual diagnostics should be interpreted carefully in the light of the estimate of the intrinsic curvature. That said, there seems to be no reason to omit the diagnostics in this case. We provide several in Figure 6.7, which is created using the following code:

```
> par(mfrow=c(1,3), mar=c(4,4,2,1), las=1)
> plot(fitted(handy.nls), residuals(handy.nls, type="pearson"),
+      xlab = "Fitted Values", ylab = "Standardized Residuals")
> abline(h=0, col="red")
> qqnorm(residuals(handy.nls, type="pearson"))
> qqline(residuals(handy.nls, type="pearson"))
> plot(fitted(handy.nls), handy.tree$dbh.cm,
+      xlab = "Fitted Values", ylab = "Observed Values")
> abline(0, 1, col="red")
```

The left frame is a plot of the standardized residuals against the fitted values, with a $y = 0$ line imposed. As with OLS regression, we want to see no particular pattern and no influential outliers. Here we seem to see a distinct kink — a systematic lack of fit in the trajectory. The center panel is a q-q plot of the residuals against the normal distribution. We prefer to see the points matching the line, and apart from one point, the match seems reasonable.

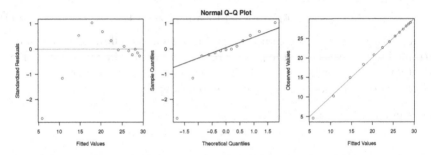

Fig. 6.7: Diagnostic plots for a simple asymptotic model fitted to the handy tree.

The right panel is a plot of the fitted values against the observed values, with an $x = y$ line imposed. We would like the points to be close to that line and not show fanning or curvature. This graphic suggests that, relative to the amount of variation being fitted by the model, the fitted model deviations are modest.

Overall, the model seems to be a pretty good fit, but we might be interested in finding a model form that accommodates the kink in the data, especially if it recurs in the trajectories of other trees.

We can learn more about the local-linearity assumptions using the `profile` function, which provides insight as to the shape of the parameter space onto which our data have been projected. Ideally we want the distribution of the parameter estimates to be approximately normal. Bates and Watts (1988) developed a graphical tool referred to as a profile-t plot, which is available in R using the following code (Figure 6.8).

```
> handy.prof <- profile(handy.nls)
> opar <- par(mfrow=c(1,2), mar=c(4,4,2,1), las=1)
> plot(handy.prof, conf = c(0.95))
```

The interpretation of the profile-t plot is as follows. These panels provide information about the acceptability of the assumption of normality on the underlying distribution of the parameter estimates. Our ideal scenario is for the solid lines to be straight and the vertical dashed lines to meet the x-axis at approximately the large-sample 95% confidence interval. However, in any case, the exact interval can be read directly from the graph or obtained using the `confint` function as we do below.

Two-dimensional likelihood profile traces are also available, as produced below using the `pairs` function (Figure 6.9). This output confirms that locally the linear approximation seems reasonable, as the traces for each parameter are close to straight. Furthermore, the parameter estimates are highly correlated because the trace lines are close to parallel. Independent parameter estimates would yield perpendicular trace lines.

```
> pairs(handy.prof)
```

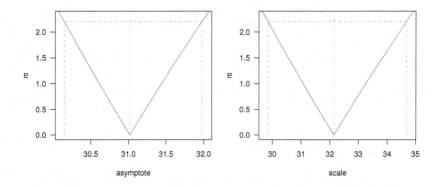

Fig. 6.8: One-dimensional profile plots for a simple asymptotic model fitted to the handy tree.

We can confirm the high correlation by calculating the estimated correlation between the parameter estimates. The scaled estimated covariance matrix is

```
> (cov.hat.mat <-
+     matrix(summary(handy.nls)$cov.unscaled, nrow=2) *
+     summary(handy.nls)$sigma^2)

           [,1]       [,2]
[1,] 0.1749164 0.4223982
[2,] 0.4223982 1.2225317
```

so the covariance is either of the off-diagonal elements

```
> cov.hat <- cov.hat.mat[1,2]
```

and the correlation is this quantity divided by the standard errors of each parameter estimate.

```
> cov.hat / prod(summary(handy.nls)$coefficients[,2])

[1] 0.913433
```

This quantity is close to one, confirming our inference about the high correlation of the parameter estimates from Figure 6.9.

6.2.5 Examining the Model

We can inspect the model object in the usual way. The t-tests against the null hypothesis that the parameters are zero are reported although they have

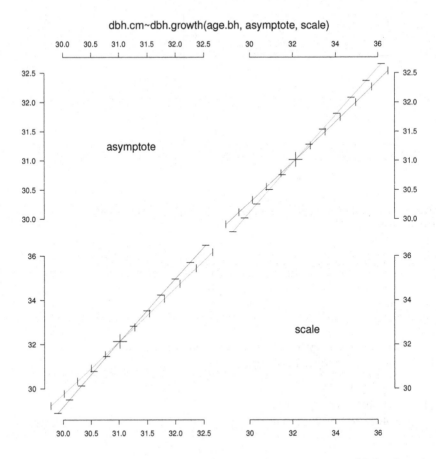

Fig. 6.9: Two-dimensional profile plots for a simple asymptotic model fitted to the handy tree.

no interpretation in this context. Interpretation of the following output is similar to that on page 185.

```
> summary(handy.nls)
```

```
Formula: dbh.cm ~ dbh.growth(age.bh, asymptote, scale)
```

```
Parameters:
          Estimate Std. Error t value Pr(>|t|)
asymptote  31.0175     0.4182   74.16 3.33e-16 ***
scale      32.1395     1.1057   29.07 9.40e-12 ***
---
Signif. codes:  0 '***' 0.001 '**' 0.01 '*' 0.05 '.' 0.1 ' ' 1
```

Residual standard error: 0.4507 on 11 degrees of freedom

Number of iterations to convergence: 5
Achieved convergence tolerance: 6.62e-06

Apart from the fact that we can't guarantee to have found a global minimum, the reported estimates are the least-squares estimates. We should try a range of other starting points to be sure that the final estimates are robust to the initial estimates.

We can compute the approximate large-sample interval estimates using

```
> (my.t <- qt(0.975, summary(handy.nls)$df[2]))
```

```
[1] 2.200985
```

```
> coef(summary(handy.nls))[,1:2] %*%
+    matrix(c(1,-my.t,1,my.t), nrow=2)
```

```
              [,1]     [,2]
asymptote 30.09697 31.93801
scale     29.70590 34.57308
```

and extract the exact marginal profiled versions using

```
> confint(handy.prof)
```

```
              2.5%    97.5%
asymptote 30.14710 31.97636
scale     29.84986 34.67562
```

These estimates are very similar, which confirms the outcomes of our diagnostics. But, the data appear to have a modest kink that our model failed to capture. This might be true just of this tree, a subset of the trees, or all the trees. We are naturally curious, but understandably reluctant to go through the same process for every tree. We will need some more powerful tools, to which we will return in Section 7.6.

Finally, we can obtain an estimate of the root-mean-squared error of the residuals using confint

```
> sqrt(sum(residuals(handy.nls)^2, na.rm=TRUE) /
+       summary(handy.nls)$df[2])
```

```
[1] 0.4507378
```

and compare it with the standard deviation of the data,

```
> sd(handy.tree$dbh.cm, na.rm=TRUE)
```

```
[1] 7.640395
```

which provides a loose indication of the predictive quality of the model. A more formal comparison requires the use of a cross-validation mechanism.

6.2.6 Using the Model

The fitted model can be used in the same manner as the fitted `lm` object: via the `predict` function. We don't need to back-transform or correct the predicted values from this model, so we don't need to write a special function for its predictions, but the following terse function simplifies the use of `curve` in Figure 6.10.

```
> handy.dbh.hat <- function(age.bh)
+    predict(handy.nls, newdata = data.frame(age.bh = age.bh))
```

We can call `handy.dbh.hat` directly as follows. Note that unless instructed otherwise, `curve` takes its range of values from the range of the current plot.

```
> par(las = 1)
> plot(dbh.cm ~ age.bh, data = handy.tree,
+        xlim = c(0, max(handy.tree$age.bh, na.rm=TRUE)),
+        ylim = c(0, max(handy.tree$dbh.cm, na.rm=TRUE)),
+        ylab = "Diameter (cm)", xlab = "Age (y)")
> curve(handy.dbh.hat, add = TRUE)
```

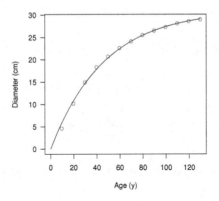

Fig. 6.10: Points and fitted line for diameter and age of a single tree from the von Guttenberg data.

6.2.7 Testing Effects

Effects in the models can be tested using the extra-sums-of-squares approach, which is also an approximate likelihood ratio test (Bates and Watts, 1988).

The test is formed along the lines of the classical F-test, computing the change in sums of squares divided by the residual sums of squares. The test statistic is asymptotically F *if the null hypothesis is true*. R provides an anova function for testing nested models that calls anovalist.nls when the arguments are nls objects.

For example, we can assess the constraint of forcing the trajectory to pass through the origin by comparing the fit of handy.nls against the self-starting asymptotic function (see Section 6.2.9). The unconstrained asymptotic function is fit by

```
> handier.nls <- nls(dbh.cm ~ SSasymp(age.bh, Asym, R0, lrc),
+                 na.action = na.exclude,
+                 data = handy.tree)
```

and the model comparison is made by

```
> anova(handy.nls, handier.nls)

Analysis of Variance Table

Model 1: dbh.cm ~ dbh.growth(age.bh, asymptote, scale)
Model 2: dbh.cm ~ SSasymp(age.bh, Asym, R0, lrc)
  Res.Df Res.Sum Sq Df Sum Sq F value    Pr(>F)
1     11    2.23481
2     10    0.36182  1  1.873  51.766 2.944e-05 ***
---
Signif. codes:  0 '***' 0.001 '**' 0.01 '*' 0.05 '.' 0.1 ' ' 1
```

The very low p-value in the outcome suggests that the unconstrained asymptotic function is a substantially better fit than the asymptotic function constrained to go through the origin that we have been using. We can verify the results of this test by checking that the interval estimate for the intercept, R0, excludes zero. The two-dimensional profile and profile-t plots (not shown here) suggest that the local-linearity assumptions are satisfactory. We compute the confidence intervals as before.

```
> confint(handier.nls)

          2.5%      97.5%
Asym 29.757344 30.575968
R0   -2.837191 -1.465437
lrc  -3.751648 -3.652848
```

The interval for R0 excludes zero by some distance, which confirms the results of the whole-model comparison.

6.2.8 Generalized Non-linear Least-Squares Models

As with linear regression, there will be circumstances in which assumptions will fail to be satisfied by our models and data. For various reasons, we will prefer to accommodate the failings within the model rather than transforming them away. The nlme package provides the gnls function, which extends nls in the same way that gls extends lm (Pinheiro and Bates, 2000). For example, to add a fitted variance model to the model of the diameter growth trajectory of the handy tree, where the variance is some power of the diameter, we would use the following code[4]:

```
> handy.gnls <-
+     gnls(dbh.cm ~ dbh.growth(age.bh, asymptote, scale),
+                   start = list(asymptote = 30, scale = 10),
+                   na.action = na.exclude,
+                   weights = varPower(form = ~age.bh),
+                   data = handy.tree)
```

Alert readers will notice that we have used 10 as the starting point for the scale parameter rather than 30, as per our earlier call to nls. This is because, at the time of writing (and for the versions of R and nlme being used[5]), the function failed to converge when the initial scale was 30. However, the function converged when the initial value was 10.

We can examine the model diagnostics and output in the usual way, although the profile and rms.curv functions are as yet unavailable.

As in Section 6.1.10, the variance function is now taken into account for all reported summaries of the model. We can use the anova function to produce a whole-model test that assesses the value of the variance function as follows.

```
> anova(handy.gnls, handy.nls)
```

	Model	df	AIC	BIC	logLik	Test	L.Ratio	p-value
handy.gnls	1	4	-12.0	-9.75	10.005			
handy.nls	2	3	20.0	21.70	-7.001	1 vs 2	34.011	<.0001

The low p-value suggests that the statistical fit of the model that includes the variance function seems substantially better than that of the model without the variance function.

---

[4] This is one of the rare occasions in which the code that we developed was platform-specific. If this code doesn't work for your platform and version of R, then we suggest that you change the starting values.

[5] 2.11.1 and 3.1-96, respectively.

6.2.9 Self-starting Functions

A handful of popular non-linear modeling functions have been retooled in R to provide their own starting values. These are called self-starting functions and are distinguished by having their names begin with the letters SS. This information is enough to be able to identify the functions in help, as typing ?SS and then TAB will provide a list of all such functions at the command prompt.[6]

We now take another brief detour into searching for R objects. We can also find all instances of these functions in the entire search path using the apropos function.

```
> apropos("^SS")
```

```
[1] "SSD"        "SSasymp"    "SSasympOff"  "SSasympOrig"
[5] "SSbiexp"    "SSfol"      "SSfpl"       "SSgompertz"
[9] "SSlogis"    "SSmicmen"   "SSweibull"
```

Note the use of the exponent symbol ^ to alert R that the string SS should start the object name. This symbol is an integral part of the so-called regular expressions (regex), which provide a very powerful and flexible framework for searching text strings.

It is possible that the returned objects are not all of the right class. We can filter all the objects in the search path by class using the following code:

```
> apropos("^SS")[sapply(apropos("^SS"),
+                  function(x) {
+                       "selfStart" %in%
+                       class(eval(parse(text = x)))}) ]
```

```
[1] "SSasymp"    "SSasympOff"  "SSasympOrig" "SSbiexp"
[5] "SSfol"      "SSfpl"       "SSgompertz"  "SSlogis"
[9] "SSmicmen"   "SSweibull"
```

NB: We can see how many objects R has in the search path using

```
> length(apropos("."))
```

```
[1] 4212
```

If the existing collection of self-starting functions is inadequate for our purposes, then we can write our own. For example, we could create an SSallometric function. R provides a selfStart function for this purpose. We need to provide a function for fitting, a function for determining initial estimates, and the names of the parameters. The function to fit will be

---

[6] This is true for all the versions of R that we use.

```
> allometric <- deriv(~ alpha * x^beta, c("alpha", "beta"),
+                     function(x, alpha, beta){},
+                     hessian = TRUE)
```

and the function for obtaining initial estimates is

```
> allometric.init <- function (mCall, data, LHS) {
+    xy <- data.frame(sortedXyData(mCall[["x"]], LHS, data))
+    if (nrow(xy) < 3)
+      stop("Too few observations to fit allometric function")
+    pars <- as.vector(coef(lm(I(log(y)) ~ I(log(x)),
+                              data = xy)))
+    pars[1] <- exp(pars[1])
+    names(pars) <- mCall[c("alpha", "beta")]
+    return(pars)
+ }
```

We then call

```
> SSallometric <- selfStart(allometric,
+                           allometric.init,
+                           c("alpha", "beta"))
```

This function then works as follows.

```
> nls(vol.m3 ~ SSallometric(dbh.cm, alpha, beta),
+     data = sweetgum)

Nonlinear regression model
  model:  vol.m3 ~ SSallometric(dbh.cm, alpha, beta)
   data:  sweetgum
     alpha        beta
0.0006338 2.1160953
 residual sum-of-squares: 2.278

Number of iterations to convergence: 5
Achieved convergence tolerance: 3.696e-06
```

6.3 Back to Maximum Likelihood

In Section 5.3.1.3, we demonstrated the use of maximum likelihood (ML) for estimating the parameters of the Weibull distribution for tree diameters. One of the advantages of ML is its flexibility. For example, we can replace any parameter in the likelihood with a function of other parameters, perhaps including predictor variables. We need to be careful about adding complexity; the ML estimates can become correspondingly harder to find, and poor choices of parameterization may really make a difference to the quality and ease of fitting of the model.

6.3.1 Linear Regression

In Section 6.1.3, we fit a linear model to predict log volume from log diameter of the sweetgum trees. We know that the parameter estimates that arise from least-squares estimation are the same as those that arise from maximum-likelihood estimation if the residuals are normally distributed. The following code fits the same regression model using maximum likelihood and provides estimates of the parameters and their asymptotic standard errors.

First, we write a function that computes the conditional, joint log-likelihood of the data. The observations are assumed to be conditionally normal. We sum the logs of the pdfs of the normal distribution evaluated at the observed data as a function of the parameter estimates.

```
> normal.ll <- function(parameters, x, y) {
+    sum(dnorm(y,
+              parameters[1] + parameters[2] * x,
+              parameters[3],
+              log = TRUE))
+ }
```

We then maximize this function using optim.

```
> good.fit <- optim(c(intercept = 1, slope = 1, sigma = 1),
+                    normal.ll,
+                    hessian = TRUE,
+                    control = list(fnscale = -1),
+                    x = sweetgum$log.dbh.cm,
+                    y = sweetgum$log.vol.m3)
```

We can now extract the maximum-likelihood parameter estimates from the returned object.

```
> good.fit$par
```

```
 intercept       slope       sigma
-7.8050519   2.2268877   0.1132759
```

The (asymptotic) estimates of the standard errors are then

```
> sqrt(diag(solve(-good.fit$hessian)))
```

```
 intercept       slope       sigma
0.14166190  0.03919263  0.01282375
```

These estimates are close to those reported previously, allowing for numerical error, although the estimate of the variance differs, as we would expect.

A disadvantage of taking this approach to fitting the linear model is that it is cumbersome to generalize in certain ways. For example, if we now wished to fit a unique slope and intercept for each species, we'd have to write that requirement into the objective function explicitly, whereas when we use lm we just add the species term to the model specification (see Section 6.1.9 for an example).

6.3.2 Non-linear Regression

A corresponding advantage is that other types of generalizations are easy. For example, we can also fit a non-linear model simply by changing the mean function.

```
> normal.ll.nl <- function(parameters, x, y) {
+    sum( dnorm(y,
+              parameters[1] * x ^ parameters[2],
+              parameters[3],
+              log = TRUE ))
+ }
```

We maximize this function using optim:

```
> good.fit <- optim(c(intercept = 1, slope = 1, sigma = 1),
+                   normal.ll.nl,
+                   hessian = TRUE,
+                   control = list(fnscale = -1),
+                   x = sweetgum$dbh.cm,
+                   y = sweetgum$vol.m3)
```

The MLEs are then

```
> good.fit$par
```

```
    intercept          slope          sigma
0.0006654548 2.1041851701 0.2411616481
```

The (asymptotic) estimates of the standard errors are

```
> sqrt(diag(solve(-good.fit$hessian)))
```

```
    intercept          slope          sigma
0.0001729483 0.0629433124 0.0272184407
```

6.3.3 Heavy-Tailed Residuals

If we are uncomfortable with the assumption that the error distribution is normal, it is a straightforward matter to choose a member of the location-scale family of the t-distribution. We will fit a linear regression with errors described by a t-distribution with 10 degrees of freedom, which provides robustness against outliers due to having heavier tails than the normal distribution.

Recall that, for any pdf $f(x)$ and constants μ and $\sigma > 0$,

$$g(x|\mu,\sigma) = \frac{1}{\sigma} f\left(\frac{x-\mu}{\sigma}\right) \tag{6.8}$$

is also a pdf (see, e.g., Casella and Berger, 1990, p. 116).

The problem with this formulation is that if $\sigma < 0$, then $g < 0$, which makes no sense in the context of g being a pdf and will create problems when we are trying to take the log of g. Therefore we want to constrain $\sigma > 0$ somehow. We could use a version of optim that supports box constraints on the parameter estimates, but it seems cleaner in this case to reparameterize the function and use $\exp(\sigma)$ as the parameter rather than σ. Now σ is unconstrained, but $g > 0$.

We can do this because maximum-likelihood estimators (MLE) are invariant under monotonic transformation. That is, if the MLE of σ is $\hat{\sigma}$, then the MLE of $\exp(\sigma)$ is $\exp(\hat{\sigma})$.

```
> t3.11 <- function(parameters, x, y) {
+    sum(dt((y - parameters[1] - x * parameters[2]) /
+           exp(parameters[3]),
+           df = 10,
+           log = TRUE) - parameters[3])
+ }
```

We again maximize this function using optim:

```
> good.fit.t <- optim(c(intercept = 1, slope = 1, sigma = 1),
+                    t3.11,
+                    hessian = TRUE,
+                    control = list(fnscale = -1),
+                    x = sweetgum$log.dbh.cm,
+                    y = sweetgum$log.vol.m3)
```

We can now extract the MLEs from the returned object.

```
> good.fit.t$par
```

```
intercept    slope      sigma
-7.770645   2.217431  -2.265762
```

The (asymptotic) estimates of the standard errors are

```
> sqrt(diag(solve(-good.fit.t$hessian)))
```

```
 intercept       slope        sigma
0.14326083 0.03943311 0.12799391
```

The interpretation of the parameters μ and σ is left to the analyst. Don't forget that the variance of the t_v is $\frac{v}{v-2}$ (Casella and Berger, 1990), so the reported scale parameter should not be directly interpreted as the antilog of the conditional standard deviation of the data. The conditional estimate of the standard deviation of the data would be

```
> exp(good.fit.t$par[3]) * sqrt(10 / (10 - 2))
```

```
   sigma
0.1159971
```

All these parameter estimators are maximum-likelihood estimators, conditional on the model, and therefore are asymptotically normal, efficient estimators, if the model is sufficiently flexible to capture the true relationship and if the assumptions hold.

Chapter 7
Fitting Linear Hierarchical Models

7.1 Introduction

We now shift to the analysis of hierarchical data using mixed-effects models. These models are a natural match for many problems that occur commonly in natural resources. A number of the tools that we discuss in this chapter are extensively documented in Pinheiro and Bates (2000), and our goal is to complement that resource, not replace it. Although we focus on mixed-effects models, other solutions are possible. For example, stochastic differential equations have also seen success in forestry (García, 1983), and generalized estimating equations may also be useful.

Recall that for fitting a linear regression using the least-squares or maximum-likelihood techniques, it was necessary to make some assumptions about the nature of the residuals or, equivalently, about the distribution of the response variable conditional on the predictor variables. The level of detail of the assumptions depended upon the applications of the model. For direct interval estimation of model parameters, it was necessary to assume that the residuals were

1. independent,
2. identically distributed, and
3. normally distributed.

It was also necessary to assume that the nature of the relationship between the response and the predictors was accurately captured by the model form. An assumption of constant variance (homoskedasticity) is implied by the "identically distributed" assumption (point 2 above).

If these assumptions are defensible, then model interpretation proceeds without complications and with a degree of comfort. However, more often than not, we will know that the assumptions are not true. This knowledge is common in natural resources data collections because the data may have a temporal structure, a spatial structure, or a hierarchical structure, or all three.[1] That structure may or may not be relevant to the scientific question, but it is very relevant to the assumptions that are necessary for data analysis and modeling.

Numerous strategies are available for modeling when the standard assumptions are unsatisfied. It may well be justifiable in some circumstances to ignore the problem altogether, depending on the application to which the model will be put. We do not advocate ignoring the problem in general.

Mixed-effects models contain both fixed and random effects. The model structure is usually suggested by the underlying design or structure of the data. An oversimplified but useful position is that random effects are suggested by the design of a study and fixed effects are suggested by the hypotheses. This position is not always true.

7.1.1 Effects

"Effects" is the label for predictor variables in a linear or non-linear model. The use of the label seems to be a hangover from experimental design and no longer really suits the application, but inertia prevents change.

The distinction between *fixed* and *random* effects can be confusing. "Random" and "fixed" are not normally held to be antonyms, or even mutually exclusive, except by sheer force of habit. Why not use "stochastic" and "deterministic"? Or "sample" and "population"? Or "local" and "global"? Such labels would provide clearer links to alternative strategies, such as stochastic differential equations. However, in order to maintain a clear connection with our source material, we shall continue to use these labels, although they appear to be both fixed, in the sense that they do not change, and random, in the sense that they lack a clear connection to their interpretation in this context.

In order to be used in a mixed-effects model, an effect has to be classified as either fixed or random. There are different ways to look at which of these two labels is best for an effect. This decision is important because the assigned label affects the data analysis and the conclusions that can be drawn. Modelers may disagree on whether effects should be fixed or random, and a predictor variable can switch from one type to the other depending on circumstances.

In some published analyses, an effect may appear to be both fixed and random in a model, for example the subplot treatment effect in the split-plot design of Pinheiro and Bates (2000), but this is merely a matter of convenience. The fixed effect that represents the treatment is numerically identical to the random effect that represents the subplot, so the same variable can be used for both roles.

---

[1] "The first law of ecology is that everything is related to everything else" (Commoner, 1971).

Statisticians have not agreed on a strategy for assigning the fixed and random roles (see, e.g., Gelman, 2005, and discussion). Some analysts claim that it depends entirely on the desired inference and some that it depends entirely on the design. In an ideal world, of course, the inference and the design are unambiguously linked. This scenario is rare in our experience.

As the statistical tools that are used to analyze such data become more sophisticated, and models that were previously unthinkable become mainstream, the inadequacies of old vocabularies become increasingly obvious. Vocabularies can affect the way we think about a problem, and inadequate vocabularies may impede progress. Robinson (1991) is excellent reading.

7.1.1.1 Fixed Effects

Fixed effects are generally held to be purposively selected, and the estimates of the levels represent only themselves. For example, if we have a fixed effect called sex with levels female and male, the statistics that we collect for level female refer only to sampling units in the population that belong to the class with level female. They are not intended to represent other, possible but unsampled, levels of sex.

In a designed experiment context, fixed effects represent the treatments, or interventions. Imagine that you have carried out an experiment and are considering repeating it. If for the experiment to be repeated it would be necessary to purposively produce the exact same levels, or even a subset of the same levels, of an experimental effect, then the effect is fixed in the design.

However, some effects that might vary upon remeasurement may also be considered fixed. An example is when the predictor variable in a regression is measured but not set. If we are interested in constructing a height–diameter equation for a particular forest and we randomly sample trees, then the diameter is usually held to be a fixed effect, even though a new sample would yield a new set of diameters. This is because the diameter of a tree has to be known in order to use the model to predict the tree's height.

Alternatively, one might say that a fixed effect is simply one for which the estimates of location (as opposed to measures of scale) are of primary interest. In the case of the height–diameter equation above, interest would lie in estimating the parameters that describe the relationship between height and diameter, as opposed to merely estimating the strength of the relationship.

Another alternative is that one might say that a fixed effect is one that the analyst wishes to condition on, for whatever reason. This pragmatic definition speaks to what we want to do with the variable rather than where it came from. That is, we classify the variable by what inference we would like to draw for it rather than how it appeared in the sample design. Hopefully, the sample design reflects the intended purpose for the variable in any case.

7.1.1.2 Random Effects

Random effects are those whose levels are supposedly sampled randomly from a range of possible levels. For example, if we have a random effect called `forest`, comprising two randomly selected forests that are represented by levels 1 and 2, then the statistics that we collect for levels 1 and 2 are intended to represent all possible levels (that is, forests) in the population from which the sample of forests was selected.

Generally, although not always, when effects are considered random, it is of interest to draw conclusions from the results of the sample *of levels* to the broader population *of levels*. That is, the levels are assumed to be collectively representative of a broader class of potential levels about which we wish to say something. In the case of the forest random effect, we might wish to make inference about the population of forests from which our sample has been drawn.

Alternatively, one might say that a random effect is simply one for which the estimates of location are not of primary interest. We might be less interested in the individual values of forest 1 and forest 2, for example, than what they can tell us about the distribution — mostly the spread — of the population of forests as a whole.

Another alternative is that one might say that a random effect is one that the analyst wishes to marginalize, for whatever reason. Again, this pragmatic definition speaks to what we want to do with the variable rather than where it came from.

In a designed experiment context, random effects represent the experimental material; that is, they identify the replication. Generally, blocks, plots, and subplots are held to be random effects.

Some authors infer from this definition of the random effect that an effect can only be random if its levels are known to be a simple random sample from a population of possible levels. We believe that this requirement is too stringent. For our purposes, an effect can be a random effect if the modeler reasonably believes the levels are representative of the population for which inference will be drawn. One way to be sure of this condition is to randomly sample the levels from the population of interest, but it is clear that other kinds of sample designs should also be acceptable.

7.1.1.3 Mixed-up Effects

Some variables do not lend themselves to easy classification, and either knowledge of the process or an epistemological slant is required. Such variables are common in natural resources. For example, if an experiment that we feel is likely to be affected by climate is repeated over a number of years, would *year* be a fixed or a random effect? It is not a random sample of possible years, but the same years would not recur if the experiment were repeated. Perhaps year might be included as a continuous fixed effect and as a categorical random effect. Likewise, there is ambiguity in classifying the replication of an experiment at known locations: some would claim that these should be a fixed effect, others that they represent environmental variation, and therefore they can be considered a random effect.

Furthermore, it may be necessary to consider as fixed effects that would otherwise be considered random, if they have only a small number of levels. Random effects are used to estimate variance components, which we describe later, and the variance components are used in inference about other terms in the model. If a variance component is poorly estimated, the effects on the rest of the model can be substantial. For example, if an experiment were split across two forests, it might seem most sensible for the forest factor to be a random effect in order that it not be necessary to condition upon the forest for model use. On the other hand, the model might be much easier to fit, and might make more sense, if the forest factor is included as a fixed effect.

7.1.2 Model Construction

The process of model construction for mixed-effects models is much more complicated than the construction of fixed-effects models. We have to balance different approaches and assumptions, each of which carries different implications for the model and its utility. If we think about the process of fitting an ordinary regression as being like a flow chart in two dimensions, then adding random effects adds a third dimension to the flow chart altogether. This additional complication is magnified by the plethora of fitting tools that can be used to fit mixed-effects models, each of which provides different functions. Therefore it is very important to plan the approach carefully before beginning.

The key point to remember is that you should be prepared to take time over this process. All the fancy graphics and scripts will amount to nothing if they are used without careful reflection. It could be argued that the increasing efficiency of model-fitting software is deleterious to the practice of statistics if it tempts the analyst to fit more models and think less about them.

The number of potential strategies is almost as varied as the number of models we can fit. Here is one that we will rely on in our further examples.

1. Choose the minimal set of fixed and random effects for the model. Specifically:

   a. Identify the random effects that must be included. These effects should be such that if they are not in the model, then the model will not adequately reflect the experimental or sample design.
   b. Identify the fixed effects that must be included. These effects should be such that if they are not in the model, then the model has no meaning.

If the data are from a designed experiment, then include all the fixed effects that will be necessary to test the hypotheses of interest.

This is the baseline model to which others will be compared.

2. Fit this model to the data using tools discussed in this chapter, and check the assumption diagnostics. Iterate through the process of refining and improving the representation of the random effects, including consideration of

   a. a heteroskedastic variance structure (several candidates),
   b. a correlation structure (several candidates), and
   c. extra random effects (e.g., random slopes).

3. When the diagnostics suggest that the fit is reasonable, consider adding more fixed effects. At each step, re-examine the diagnostics to be sure that any estimates that you will use to assess the fixed effects are based on a good match between the data, model, and assumptions.

We note in passing that often the software that we use to fit models constrains the range of models that can be fit with finite effort.

A further layer of complexity is that it may well be that some important assumptions cannot be met in the absence of certain fixed effects or random effects. That is, a satisfactory resolution of step 3 above may never be achieved because the key fixed effects are missing. In this case, a certain amount of iteration is inevitable and careful record-keeping is *essential*. And compromise might be necessary.

The roles of the fixed and the random effects are distinct in the tools that we will describe. Fixed effects *explain* variation. Random effects *organize the unexplained* variation. The careless analysis of certain experimental designs may result in error. Be sure to check whether the between-block variation is being used according to the experimental design.

At the end of the model-fitting process, you will have a model that may superficially seem worse than a simple linear regression by most metrics of model quality, for example residual variance. Adding random effects may add information, and may improve diagnostic compatibility, but does not explain more variation!

The bottom line is that the goal of the analyst is to find the simplest model that satisfies the necessary model assumptions and answers the questions of interest. It is tempting to go hunting for more complex random effects structures, which may provide a higher maximum likelihood, but if the simple model satisfies the assumptions and answers the questions, then trying to maximize the likelihood further may not bear fruit. Schabenberger and Pierce (2002) provide some wonderfully practical advice: "Don't be afraid to start, and don't be afraid to finish".

7.1.3 Solving a Dilemma

We approach the advantages to modeling that are offered by mixed-effects models through a simple example. Imagine that we are interested in constructing a height–diameter relationship using two randomly selected plots in a forest and that we have measured three trees on each plot. It turns out that on the plots the growing conditions are quite different, leading to a systematic but unexpected difference between the height–diameter relationships on each. We plot the data in Figure 7.1 using the following code:

```
> trees <- data.frame(plot=factor(c(1, 1, 1, 2, 2, 2)),
+                       dbh.cm=c(30, 32, 35, 30, 33, 35),
+                       ht.m=c(25, 30, 40, 30, 40, 50))

> plot(trees$dbh.cm, trees$ht.m, pch=c(1, 19)[trees$plot],
+       xlab="Diameter (cm)", ylab="Height (m)")
> abline(lm(ht.m ~ dbh.cm, data=trees), col="darkgrey")
```

If we fit a simple regression to the trees, then we obtain a residual/fitted value plot as displayed in Figure 7.2.

```
> case.model.1 <- lm(ht.m ~ dbh.cm, data=trees)
> plot(fitted(case.model.1), residuals(case.model.1),
+       ylab = "Residuals", xlab = "Fitted Values",
+       pch = c(1, 19)[trees$plot])
> abline(h = 0, col = "darkgrey")
```

If we fit a simple regression to the trees with an intercept for each plot, then we obtain a residual/fitted value plot as displayed in Figure 7.3.

```
> case.model.2 <- lm(ht.m ~ dbh.cm*plot, data=trees)
> plot(fitted(case.model.2), residuals(case.model.2),
+       ylab = "Residuals", xlab = "Fitted Values",
+       pch = c(1, 19)[trees$plot])
> abline(h = 0, col = "darkgrey")
```

Figures 7.1–7.3 show the analyst's dilemma. The residuals in Figure 7.2 clearly show a correlation structure within the plots; the plot conditions dictate that all the trees in the plot will have the same sign of residual. This phenomenon is not seen in Figure 7.3. However, the model described by Figure 7.3 has limited utility because in order to use it for an unmeasured tree we have to nominate whether the unmeasured tree belongs to plot 1 or plot 2. If the tree belongs to neither plot, which is true of all of the unmeasured trees, then the model can make no prediction. So, the dilemma is that we can construct a useless model that satisfies the regression assumptions or a useful model that definitely does not.

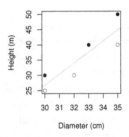
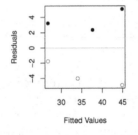
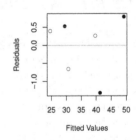

Fig. 7.1: Height and diameter for three trees on two plots (full and outline symbols) with a line of least–squares regression.

Fig. 7.2: Residual plot for height–diameter model for three trees on two plots (full and outline symbols).

Fig. 7.3: Residual plot for height–diameter model including plot for three trees on two plots (full and outline symbols).

7.1.4 Decomposition

The dilemma documented in Section 7.1.3 has several solutions. One is to use mixed-effects models, and the others, which we do not cover here, are explicit modeling of the correlation structure using generalized least squares, generalized estimating equations, etc.

The mixed-effects models approach is to decompose the unknown variation into smaller pieces, each of which satisfies decomposed versions of the necessary assumptions. Imagine that we could take the six residual values presented in Figure 7.2, which have the plot-level correlation structure, and decompose each of them into two plot-level errors and six within-plot errors. That is, instead of symbolizing the difference between the observations and the predictions using six residuals, we could use eight. Instead of

$$y_{ij} - \hat{y}_{ij} = \hat{\varepsilon}_{ij} \tag{7.1}$$

with $\varepsilon_{ij} \sim \mathcal{N}(0, \sigma^2)$, we could write

$$y_{ij} - \hat{y}_{ij} = \hat{b}_i + \hat{\varepsilon}_{ij} \tag{7.2}$$

Then we would need to assume that:

- The true relationship between x and y is linear.
- $b_i \sim \mathcal{N}(0, \sigma_b^2)$.
- $\varepsilon_{ij} \sim \mathcal{N}(0, \sigma^2)$.
- All the ε_{ij}s are independent.
- The sample represents the population for which inference is being made.

This collection of assumptions is arguably better for our circumstances. Furthermore, when the time comes to use the model for prediction, we do not need to know the plot label, as the fixed effects do not require that we know it. The plot-to-plot variation is represented by σ_b^2 and no longer creates a correlation structure within the residuals. Also, although it seems odd to use eight quantities to represent six unknowns, the fact that we put constraints on them (for example, zero mean) means that the quantities can be estimated, although they do not need to be estimated in order to fit the model.

We now have two unknown variances in our model. These variances are generally referred to as variance components, inasmuch as they represent the sources of variation that contribute to the errors in the model.

This example illustrates the use of random effects. Random effects do not explain variation. Explaining variation is the role of the fixed effects. Random effects organize variation, or enforce a more complex structure upon it, in such a way that a match is possible between the model assumptions and the diagnostics. We would expect the overall uncertainty, measured as root-mean-squared error, to increase any time we estimate parameters for a model using any way other than by least squares.

7.2 Linear Mixed-Effects Models

Now we use a simple example as a basis for walking through the code to fit mixed-effects models in R.

7.2.1 A Simple Example

We start with a very simple and abstract example. Our goal is to construct a model to predict y from a continuous predictor variable x and a categorical predictor variable called group. First we generate a simple dataset.

```
> example <- data.frame(y = c(4.2, 4.8, 5.8, 1.2, 10.1,
+                             14.9, 15.9, 13.1),
+                       x = c(1, 2, 3, 4, 1, 2, 3, 4),
+                       group = factor(c(1, 1, 1, 1,
+                                        2, 2, 2, 2)))
```

Now, we have to load the package that holds the mixed-effects code, nlme (Pinheiro and Bates, 2000).

```
> library(nlme)
```

Next, we plot the data (Figure 7.4).

```
> par(las=1, mar=c(4,4,1,1))
> colours <- c("red", "blue")
```

```
> plot(y ~ x, data = example,
+       col = colours[group],
+       pch = as.numeric(group))
```

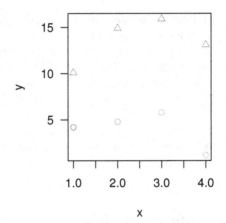

Fig. 7.4: A simple dataset to show the use of mixed-effects models. Circles are from group 1 and triangles are from group 2.

7.2.1.1 Linear Regression

We provide only modest information about linear regression as it is covered more thoroughly in Chapter 6. This model is just trying to predict y using x. In algebraic notation, we would write

$$y_i = \beta_0 + \beta_1 \times x_i + \varepsilon_i \qquad (7.3)$$

where β_0 and β_1 are fixed but unknown population parameters and ε_i are residuals. The following assumptions are required:

1. The true relationship between x and y is linear.
2. $\varepsilon_i \sim \mathcal{N}(0, \sigma^2)$.
3. ε_i are independent.
4. The sample represents the population for which inference is being made.

Note that the values of β_0 and β_1 that minimize the residual sum of squares are the least-squares estimates in any case and are unbiased if assumption 1 is true. If assumptions 2 and 3 are also true, then the estimates also have other desirable properties: they are minimum-variance, unbiased estimators.

The model is fit using R with the following code:

```
> basic.1 <- lm(y ~ x, data=example)
```

We can examine estimates from the model via:

```
> coef(summary(basic.1))
```

	Estimate	Std. Error	t value	Pr(>|t|)
(Intercept)	8.5	5.142956	1.65274590	0.1494694
x	0.1	1.877942	0.05324978	0.9592616

Next, we let each **group** have its own intercept. In algebra,

$$y_i = \beta_{01} + \beta_{02} \times g_i + \beta_1 \times x_i + \varepsilon_i \tag{7.4}$$

where β_{01}, β_{02}, and β_1 are fixed but unknown population parameters, g_i is an indicator variable with value 0 for group 1 and 1 for group 2, and ε_i are residuals. Note that in accordance with R's convention, β_{01} is the coefficient representing the intercept for the first group and β_{02} is the coefficient representing the difference between the intercepts for the first and the second groups. The same assumptions are required as for model (7.3).

The model is fit using R with the following code:

```
> basic.2 <- lm(y ~ x + group, data=example)
```

Now we let each **group** have its own intercept and slope.

$$y_i = \beta_{01} + \beta_{02} \times g_i + (\beta_{11} + \beta_{12} \times g_i) \times x_i + \varepsilon_i \tag{7.5}$$

where β_{01}, β_{02}, β_{12}, and β_{12} are fixed but unknown population parameters, g_i is an indicator variable with value 0 for group 1 and 1 for group 2, and the ε_i are the errors. The same assumptions are required as for model (7.3).

The model is fit using R with the following code:

```
> basic.3 <- lm(y ~ x * group, data=example)
```

7.2.1.2 Mixed Effects

In order to analyze the data using a mixed-effects model, some preparation is advantageous. It will help to convert the data to a grouped object — a special kind of data frame that supports special **nlme** commands. The **group** will hereby be a random effect. Constructing a **groupedData** object is not essential, but it enables functionality later on. For example, we will be able to use **augPred**, as seen in Figure 7.5, below.

```
> example.mixed <- groupedData(y ~ x | group, data=example)
```

Now we fit the basic mixed-effects model that allows the intercepts to vary randomly between the groups. This is analogous to the situation described in the previous section. We will add a subscript for clarity. The model form is

$$y_{ij} = \beta_0 + b_{0i} + \beta_1 \times x_{ij} + \varepsilon_{ij} \qquad (7.6)$$

where β_0 and β_1 are fixed but unknown population parameters, the b_{0i} are the two group-specific random intercepts, and ε_{ij} are residuals. The following assumptions are required:

- The true relationship between x and y is linear.
- $b_{0i} \sim \mathcal{N}(0, \sigma_{b_0}^2)$.
- $\varepsilon_{ij} \sim \mathcal{N}(0, \sigma^2)$.
- The b_{0i} and the ε_{ij} are all independent.
- The samples represent the populations from which they were drawn.

The model is fit using R with the following code:

```
> basic.4 <- lme(y ~ x,
+                 random = ~1 | group,
+                 data = example.mixed)
```

The `random` syntax can be a little confusing. Here we are instructing R to let each group have its own random intercept. If we wanted to let each group have its own slope and intercept, we would write `random = ~x | group`. If we wanted to let each group have its own slope but only a common intercept, we would write `random = ~x - 1 | group`. We can also write the model in list form, that is, `random = list(group = ~ 1)`.

We can examine the model in a useful graphic called an *augmented pre-diction plot*. This plot provides a scatterplot of the data, with a unique panel for each group, and a fitted line that represents the model predictions (Figure 7.5). We should also check the regression diagnostics that are relevant to our assumptions, but we have so few data here that the diagnostics are not instructive. We will develop these ideas further during the case study that follows.

```
> plot(augPred(basic.4))
```

If we are satisfied with the model diagnostics, then we can examine the structure of the model, including the estimates, using the `summary` function. The summary function presents a collection of useful information about the model. Here we report the default structure for a `summary.lme` object; that is, the object produced when the `summary` function is called on an object of type `lme`.

```
> summary(basic.4)
```

First, the `data.frame` object is identified and fit statistics are reported, including Akaike's information criterion, Schwartz's Bayesian information criterion, and the log-likelihood.

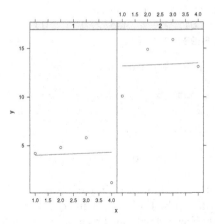

Fig. 7.5: An augmented plot of the basic mixed-effects model with random intercepts fit to the sample dataset.

```
Linear mixed-effects model fit by REML
 Data: example.mixed
       AIC      BIC    logLik
  43.74387 42.91091 -17.87193
```

The random-effects structure is then described, and estimates are provided for the variance components. Here we have an intercept for each group for which the standard deviation is reported as well as the standard deviation of the residuals within each group.

```
Random effects:
 Formula: ~1 | group
         (Intercept) Residual
StdDev:     6.600769 2.493992
```

The fixed effects structure is described next in a standard t-table arrangement. Estimated correlations between the fixed effects follow.

```
Fixed effects: y ~ x
             Value Std.Error DF   t-value p-value
(Intercept)    8.5  5.142963  5 1.6527437  0.1593
x              0.1  0.788670  5 0.1267958  0.9040
 Correlation:
   (Intr)
x -0.383
```

The distribution of the within-group residuals, also called the *innermost residuals* in the context of strictly hierarchical models by Pinheiro and Bates (2000), is then described.

Standardized Within-Group Residuals:

Min	Q1	Med	Q3	Max
-1.2484738	-0.4255493	0.1749470	0.6387985	1.0078956

Finally, the hierarchical structure of the model and data is presented.

```
Number of Observations: 8
Number of Groups: 2
```

7.2.1.3 Mixed Effects, Unique Variances

Next we shall allow the residuals within each group to have their own variances. The model form will be the same as in equation (7.6), but the assumptions will be different. Now we will need to assume that the residuals ε_{ij} from each group have their own distributions. That is:

- The true relationship between x and y is linear.
- $b_{0i} \sim \mathcal{N}(0, \sigma^2_{b_1})$.
- $\varepsilon_{1j} \sim \mathcal{N}(0, \sigma^2_{b01})$.
- $\varepsilon_{2j} \sim \mathcal{N}(0, \sigma^2_{b02})$.
- The b_{0i} and the ε_{ij} are all independent.
- The samples represent the populations from which they were drawn.

The model is fitted using R with the following code:

```
> basic.5 <- lme(y ~ x, random = ~1 | group,
+              weights = varIdent(form = ~1 | group),
+              data = example.mixed)
```

The summary output is essentially identical to the previous output in structure, with the addition of a new section that summarizes the newly added variance model. Here we show only the new portion.

```
> summary(basic.5)
```

```
Variance function:
 Structure: Different standard deviations per stratum
 Formula: ~1 | group
 Parameter estimates:
        1        2
 1.000000 1.327843
```

We see that the standard deviation of the first group has been set to 1 and the standard deviation of the other groups is presented as a ratio relative to group 1.

7.2.1.4 Mixed Effects, Unique Variances, Autocorrelation

Finally, we allow for simple temporal autocorrelation within each group. We will specify a one-step autocorrelation that specifies that the correlation between any pair of residuals is a function of their distance apart in time (or space). Again, the model form will be the same as in equation (7.6), but the assumptions will be different. Now, we will need to assume that:

- The true relationship between x and y is linear.
- $b_{0i} \sim \mathcal{N}(0, \sigma_{b_1}^2)$.
- $\varepsilon_{1j} \sim \mathcal{N}(0, \sigma_{b01}^2)$.
- $\varepsilon_{2j} \sim \mathcal{N}(0, \sigma_{b02}^2)$.
- $\mathrm{Corr}(\varepsilon_{ab}, \varepsilon_{ac}) = \rho^{|c-b|}$.
- The b_{0i} and the ε_{ij} are all independent otherwise.
- The samples represent the populations from which they were drawn.

The model is fit using R with the following code:

```
> basic.6 <- lme(y ~ x, random = ~1 | group,
+                weights = varIdent(form = ~1 | group),
+                correlation = corAR1(),
+                data = example.mixed)
```

The summary output is again essentially identical to the previous output in structure, with the addition of a new section that summarizes the newly added correlation model. Here we show only the new portion.

```
> summary(basic.6)

Correlation Structure: AR(1)
 Formula: ~1 | group
 Parameter estimate(s):
      Phi
0.8107325
```

The model estimates the first-order autocorrelation between the residuals within the groups as 0.81, which is quite high.

Any or all of the innovations that we added to the model in the recent fitting exercise might be necessary to achieve a suitable model fit.

7.3 Case Study: Height and Diameter Model

We now use a real dataset to demonstrate the construction of mixed-effects models. Section 2.4.5 summarizes the data entry and processing. Briefly, a sample of 66 trees was purposively selected in national forests around northern and central Idaho. The habitat type and diameter at 1.37 m $(4'6'')$ were also recorded for each tree, as was the national forest from which it came. Each tree was then split, and decadal measures were made of height and diameter inside bark at 1.37 m ($4'6''$). We have data from nine national forests and six different habitat types.

```
> names(stage)

 [1] "Tree.ID"   "Forest"    "HabType"   "Decade"    "Dbhib"
 [6] "Height"    "Age"       "Forest.ID" "Hab.ID"    "dbhib.cm"
[11] "height.m"

> dim(stage)

[1] 542  11
```

7.3.1 Height vs. Diameter

The prediction of height from diameter provides useful and inexpensive information. It may be that the height vs. diameter relationship differs among habitat types, climate zones, or tree ages. We shall examine the height/diameter model of the trees using a mixed-effects model. We'll start with a simple case, using only the oldest measurement from each tree.

```
> stage.old <- stage[stage$Decade == 0, ]
```

Note that this code actually drops a tree from our dataset because that tree lacks a measurement at decade 0, but we can afford to let it go for the purposes of this demonstration.

Based on our knowledge of the locations of national forests, it seems reasonable to believe that there will be similarities between trees that grow in the same forest *relative to the overall population of trees*. That is, a randomly selected pair of trees from the same forest are more likely to be similar to one another than a randomly selected pair of trees from the whole population, which includes all the different forests. However, we would like to create a model that does not need to rely on knowing the national forest; that is, a model that can plausibly be used for trees in other forests. The following approach is acceptable as long as we are willing to believe that the sample of trees that we are using is representative of the conditions for which we wish to apply the model. In the absence of other information, this decision is a judgment call. We assume it for the moment.

So, based on the information above, national forest will be a random effect and habitat type will be a fixed effect, the inclusion of which we will test for. That is, we wish to construct a model that can be used for any forest, and that might be more accurate if used correctly within a named national forest, and provides unique estimates for habitat type. We can later ask how useful the knowledge of habitat type is and whether we want to include it in the model. So, we'll have two random effects: national forest and tree within national forest. We have one baseline fixed effect, diameter at breast height inside bark, with two potential additions, age and habitat type. The lowest-level sampling unit will be the tree, nested within national forest.

As noted earlier, it is convenient to provide a basic `groupedData` structure to R. The structure will help R create useful graphical diagnostics later in the analysis.

```
> stage.old <- groupedData(height.m ~ dbhib.cm | Forest.ID,
+                          data = stage.old)
```

Now, we look to our model. An algebraic expression of the model is

$$y_{ij} = \beta_0 + b_{0i} + \beta_1 \times x_{ij} + \varepsilon_{ij} \qquad (7.7)$$

where y_{ij} is the height of tree j in forest i and x_{ij} is the diameter of the same tree. β_0 and β_1 are fixed but unknown parameters, and b_{0i} are the forest-specific random and unknown intercepts. Later we might see if the slope also varies with forest. So, in matrix form from Laird and Ware (1982), the model is

$$\mathbf{Y} = \mathbf{X}\boldsymbol{\beta} + \mathbf{Z}\mathbf{b} + \boldsymbol{\varepsilon}$$
$$\mathbf{b} \sim \mathcal{N}(\mathbf{0}, \mathbf{D})$$
$$\boldsymbol{\varepsilon} \sim \mathcal{N}(\mathbf{0}, \mathbf{R})$$

where Y is the column of tree heights and X is the column of diameters bound with a column of 1s for the intercept. β will be a vector of parameter estimates. Z will be a matrix of 0s and 1s to allocate the observations to different forests. b will be a vector of means for the forests and trees within forests. Finally, we will let $\mathbf{D}$ be a 9×9 identity matrix multiplied by a constant σ_h^2, as there are nine national forests, and $\mathbf{R}$ be a 66×66 identity matrix multiplied by a constant σ^2, as there are 66 trees.

NB: $\mathbf{D}$ and $\mathbf{R}$ are covariance matrices constructed using a small number of parameters. The structure of $\mathbf{D}$ and $\mathbf{R}$ is *suggested* by what is known about the data and can be *tested* by comparing nested models.

The key assumptions that we're making for our model are that:

1. The model structure is correctly specified.
2. The random effects are normally distributed.
3. The innermost residuals are normally distributed.
4. The innermost residuals are homoskedastic within and across the groups.
5. The innermost residuals are independent within the groups.
6. The sample represents the population for which inference is being made.

It is timely to introduce some useful nomenclature. For hierarchical models, there is more than one level of fitted values and residuals. Pinheiro and Bates (2000) adopt the following labels: the *outermost* residuals and fitted values are conditional only on the fixed effects, the *innermost* residuals and fitted values are conditional on the fixed and all the random effects, and there are as many levels between these extremes as necessary. So, in a two-level model like this one:

- The outermost residuals are the residuals that are computed from the outermost fitted values, which are computed using only the fixed effects. Let's refer to them as r_0.

$$r_0 = y_{ij} - \hat{\beta}_0 - \hat{\beta}_1 \times x_{ij} \qquad (7.8)$$

- The innermost residuals are the residuals that are computed from the innermost fitted values, which are computed from the fixed effects and the random effects. We shall refer to them as r_1.

$$r_1 = y_{ij} - \hat{\beta}_0 - \hat{b}_{0i} - \hat{\beta}_1 \times x_{ij} \qquad (7.9)$$

Note that in the more general case of mixed-effects models with crossed random effects, the labels "innermost" and "outermost" are not likely to be useful.

The linear mixed-effects apparatus also provides us with three kinds of innermost and outermost residuals:

1. *response* residuals, simply the difference between the observation and the prediction;
2. *Pearson* residuals, which are the response residuals scaled by dividing by their estimated standard deviation; and
3. *normalized* residuals, which are the Pearson residuals pre-multiplied by the inverse square root of the estimated correlation matrix from the model.

In the model specification, we are not making any assumptions about the outermost residuals. However, they are useful for summarizing the elements of model performance. We fit a model to the measurements taken at the latest point for each tree as follows. First, we construct the `groupedData` object.

```
> stage.old <- groupedData(height.m ~ dbhib.cm | Forest.ID,
+                          data = stage[stage$Decade == 0,])
```

Our model is then fit using the following code:

```
> hd.lme.1 <- lme(height.m ~ dbhib.cm,
+                 random = ~1 | Forest.ID,
+                 data = stage.old)
```

We next construct diagnostic graphs to check our assumptions. Note that in some cases the assumptions are stated in an unhelpfully broad fashion.

Therefore the sensible strategy is to check for the conditions that can be interpreted in the context of the design, the data, and the incumbent model. For example, there are infinite ways that the innermost residuals could fail to have constant variance. We should ask: What are the important ways? The situation that is most likely to lead to problems is if the variance of the residuals is a function of some effect, whether that be a fixed effect or a random effect.

Rather than trust our ability to anticipate what the programmers meant by the labels that they use, etc, we prefer to know what goes into each of our plots. The best way to do that is to put it there ourselves. To examine each of the assumptions in turn, we have constructed the following suite of graphics. These are presented in Figure 7.6.

1. A plot of the outermost fitted values against the observed values of the response variable. This graph allows an overall summary of the explanatory power of the model.

```
> scatter.smooth(fitted(hd.lme.1, level=0),
+                 stage.old$height.m,
+                 xlab = "Fitted Values (height, m.)",
+                 ylab = "Observed Values (height, m.)",
+                 main = "Model Structure (I)")
> abline(0, 1, col = "blue")
```

The important questions to ask are:

a. How much of the variation is explained by the fixed effects?
b. How much of the variation remains?
c. Is there evidence of lack of fit anywhere in particular?

2. A plot of the innermost fitted values against the innermost Pearson residuals. This graph allows a check of the assumption of correct model structure.

```
> scatter.smooth(fitted(hd.lme.1),
+                 residuals(hd.lme.1, type="pearson"),
+                 main = "Model Structure (II)",
+                 xlab = "Fitted Values",
+                 ylab = "Innermost Residuals")
> abline(h = 0, col = "red")
```

The important questions to ask are:

a. Is there curvature? If so, then perhaps the fixed effects could be augmented or improved. There might be an interaction missing, or perhaps a predictor should be quadratic.
b. Do the residuals fan out? If so, then the variance function for the model could be inadequate. Consider modeling the change in variance explicitly. Or, if this is coupled with curvature and a biological explanation, a transformation might be useful.

c. Are any outliers evident? If so, then they should be checked and their influence on the model assessed by refitting the model without them.

3. A q-q plot of the estimated random effects to check whether they are normally distributed *with constant variance*. Note that interpretation of the q-q plot is sensitive to the assumption of constant variance, so violations of this diagnostic could be due to non-normality *or heteroskedasticity*.

```
> ref.forest <- ranef(hd.lme.1)[[1]]
> ref.var.forest <-
+    tapply(residuals(hd.lme.1, type="pearson", level=1),
+          stage.old$Forest.ID,  var)
> qqnorm(ref.forest, main="Q-Q Norm: Forest Random Effects")
> qqline(ref.forest, col="red")
```

The important questions to ask are:

a. Do the points follow a straight line, or do they exhibit patterns that can be translated as skewness or kurtosis? If the latter, then the variance function for the model could be inadequate. Consider modeling the change in variance explicitly. Also, a transformation *may* be required, but care should be taken. Perhaps large-sample theory can be invoked.

b. Are any outliers evident? See possible actions in part 2c above. The identification of outlying groups can be particularly informative.

4. A q-q plot of the Pearson residuals to check whether they are normally distributed with constant variance.

```
> qqnorm(residuals(hd.lme.1, type="pearson"),
+        main="Q-Q Normal - Residuals")
> qqline(residuals(hd.lme.1, type="pearson"), col="red")
```

The important questions to ask are:

a. Do the points follow a straight line, or do they exhibit skewness or kurtosis? See possible actions in part 3a above.

b. Are any outliers evident? See possible actions in part 2c above.

5. A notched boxplot of the innermost Pearson residuals by the grouping variable to see what the within-group distribution looks like.

```
> boxplot(residuals(hd.lme.1, type = "pearson", level = 1) ~
+         stage.old$Forest.ID,
+         ylab = "Innermost Residuals",
+         xlab = "National Forest",
+         notch = TRUE,
+         varwidth = TRUE,
+         at = rank(ref.forest))
> axis(3, labels = format(ref.forest, dig=2),
```

```
+        cex.axis = 0.8, at = rank(ref.forest))
> abline(h = 0, col = "darkgreen")
```

The important questions to ask are:

a. Do the notches intersect 0? If not, then the relevant groups should be checked and their effect on the model fit examined by deleting them.
b. Is there a trend between the medians of the within-group residuals and the estimated random effect? If so, then the random effect may be explaining some variation that could be explained by a fixed effect.

6. A scatterplot of the variance of the Pearson residuals within the forest against the forest random effect.

```
> plot(ref.forest, ref.var.forest,
+        xlab = "Forest Random Effect",
+        ylab = "Variance of within-Forest Residuals")
> abline(lm(ref.var.forest ~ ref.forest), col="purple")
```

a. Is there a distinct positive or negative trend? If so, then careful thought about the variance model might be beneficial.

Of course, there is no need to pack all the graphical diagnostics into one figure. We cross-reference the list against Figure 7.6. In fact, all of these residual diagnostics look good for the model that we have fit.

The next important question is whether there are any outliers or high-influence points. In a case like this, it is relatively easy to see from the diagnostics that no point is likely to dominate the fit in this way. However, a more formal examination of the question may be valuable. There is considerable peer-reviewed development of the problem of outlier and influence detection in mixed-effects models, reaching back at least to Christensen et al. (1992). Schabenberger (2005) provides an overview of the extensive offerings available in SAS, none of which are presently available in R packages as far as we know. Demidenko (2004) and Demidenko and Stukel (2005) also suggest some alternatives.

It is worth recalling that much of the outlier-detecting apparatus focuses on efficiently assessing the effect of individual observations on the model. Therefore, if the fitting process is reasonably quick and the analyst is patient, it is straightforward to custom-build influence diagnostics that focus specifically upon the parameter estimates or tests of interest. The simplest strategy is to refit the model, dropping each observation or group one by one and collecting the results in a vector for further analysis. This is easily handled by using the **update()** function, and we do not cover it here.

We will accept the model as it stands for the moment and go on to examine the model summary. We divide the summary into pieces here to explain its structure and content.

```
> summary(hd.lme.1)
```

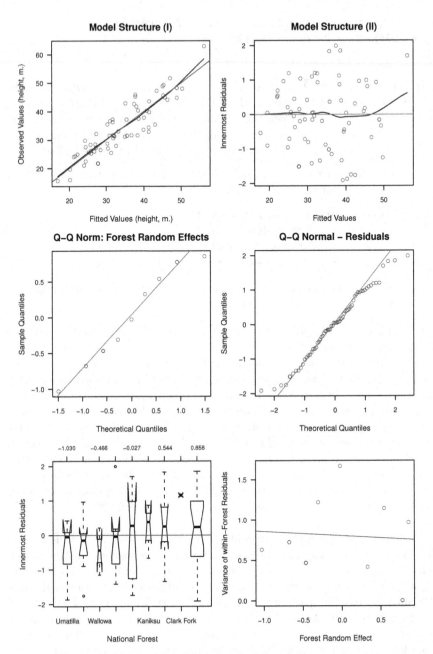

Fig. 7.6: Selected diagnostics for the fit of the mixed-effects model of height against diameter, with each national forest having a random intercept, for Stage's data (`hd.lme.1`).

1. Linear mixed-effects model fit by REML
    Data: stage.old
           AIC   BIC   logLik
        376.6805 385.2530 -184.3403

    Here we have the overall metrics of model fit, including the log-likelihood (recall that this is the quantity that we are maximizing to make the fit) and the AIC and BIC statistics. The fixed effects are profiled out of the log-likelihood, so that the log-likelihood is a function only of the data and two parameters: σ_h^2 and σ^2. We are not comparing these statistics with anything, so they do not offer direct interpretation here.

2. Random effects:
    Formula: ~1 | Forest.ID
                 (Intercept) Residual
    StdDev: 1.151405   3.937486

    The formula reminds us of what we asked for: that the forest be a random effect and that a unique intercept be fit for each level of forest. The square roots of the estimates of the two parameters are also here. These quantities are measured in *meters*. That is, the standard deviation of the variation between the forest mean heights is 1.15 m and that of the tree heights within the forests is 3.94 m.

3. Another metric of model quality is RMSE, which is the estimate of the standard deviation of the response residuals conditional on only the fixed effects. Note that 3.94 is *not* the RMSE but it is instead an estimate of the standard deviation of the response residuals conditional on the fixed and the random effects. Obtaining the RMSE is relatively easy because the random effects and the residuals are assumed to be independent.

$$\text{RMSE} = \sqrt{\sigma_h^2 + \sigma^2} = 4.1$$

The last metric of model quality we can get here is the intra-class correlation. This is the variance of the random effect divided by the sum of the variances of the random effects and the residuals

$$\rho = \frac{\sigma_h^2}{\sigma_h^2 + \sigma^2} = 0.0788$$

so about 7.9% of the variation in height (that is not explained by diameter) is explained by national forest. This is a small quantity and implies that the systematic differences between tree heights in the different forests are small once the tree diameter has been taken into account.

4. Fixed effects: height.m ~ dbhib.cm

	Value	Std.Error	DF	t-value	p-value
(Intercept)	6.58239	1.7763571	55	3.705556	5e-04
dbhib.cm	0.57036	0.0335347	55	17.008062	0e+00

Now we have a reminder of the fixed-effects model and the estimates of the fixed effects. We have several columns:

a. the value of the estimate,
b. its standard error,
c. the degrees of freedom,
d. the t-value associated with the significance test of the null hypothesis that the estimate is 0 against the two-tailed alternative that it is not 0, which is meaningless for this particular model, and
e. the p-value associated with that meaningless test.

5. Correlation:
```
        (Intr)
dbhib.cm -0.931
```

This is the correlation matrix for the estimates of the fixed effects. It is estimated from the design matrix. This comes from the covariance matrix of the fixed effects, which can be estimated by $(\mathbf{X}'\hat{\mathbf{V}}^{-1}\mathbf{X})^{-1}$. Here the correlation is high and negative, which in our case is simply a consequence of not centering the data before fitting the model. The high negative correlation is to be expected and is not of great diagnostic importance in this case.

6. Standardized Within-Group Residuals:
```
      Min          Q1          Med          Q3          Max
-1.91215622 -0.70233393  0.04308139  0.81189065  1.99133843
```

This segment reports distributional information about the within-group residuals. We should be asking ourselves: Are they symmetric? Are there egregious outliers? We can compare these values to what we know of the standard normal distribution, for which the median should be about 0, the first quartile at -0.674, and the third quartile at 0.674.

7. Number of Observations: 65
Number of Groups: 9

And finally, we have confirmation that we have the correct number of observations and groups. This is a useful conclusion to draw; it comforts us that we fit the model that we thought we had!

A set of sequential tests of the model's fixed effects can be had from

```
> anova(hd.lme.1)
```

```
              numDF denDF  F-value  p-value
(Intercept)       1    55 2848.4436  <.0001
dbhib.cm          1    55  289.2742  <.0001
```

Pinheiro and Bates (2000) recommend that these tests be used in preference to whole-model tests and Wald tests, which are reported in item 4 above. It is worth noting at this point that the question of testing in the more general case of models with crossed random effects is not satisfactorily resolved. The `anova.lme` function, that is, the version of `anova` that is used when the first argument is of class lme, also has a `type` argument, for which "marginal" is an option and that defaults to "sequential".

7.3.2 Use More Data

We now treat the grand fir height/diameter data from Stage (1963) in a different way. We have numerous measurements of height and diameter for each tree. It seems wasteful to use only the oldest observation.

We still assume that the national forests represent different purposively selected sources of climatic variation and that habitat type represents a randomly selected treatment of environment (it is probably not true, but we assume that it is). This is like a randomized block design, where the blocks and the treatment effects are crossed. This time we are interested in using all of the data. Previously we took only the first measurement. How will the model change? We begin by setting up the data in a `groupedData` object.

```
> stage <- groupedData(height.m ~ dbhib.cm | Forest.ID/Tree.ID,
+                      data = stage)
```

We say that, based on the information above, national forest will be a random effect and habitat type a candidate fixed effect. So, we will have one to three fixed effects; dbhib, age, and habitat, and two random effects; forest and tree within forest. The response variable will still be the height measurement. There will be numerous height measurements within each tree, separated by time. There will be numerous trees measured within each forest, separated by space. We assume, for the moment, that the measurements are conditionally independent within the tree. This means that we are assuming that the innermost residuals from the model that we have just described will be independent of one another. This assumption is definitely not true, and will be revisited. Now, we look to our model.

$$y_{ijk} = \beta_0 + b_{0i} + b_{0ij} + \beta_1 \times x_{ijk} + \varepsilon_{ijk} \qquad (7.10)$$

where y_{ijk} is the height of tree j in forest i at measurement k and x_{ijk} is the diameter of the same tree. β_0 and β_1 are fixed but unknown parameters, b_{0i} are the forest-specific random and unknown intercepts, and b_{0ij} are the tree-specific random and unknown intercepts. Later we might see if the slope also varies with forest. So, in matrix form, we have

$$\mathbf{Y} = \mathbf{X}\boldsymbol{\beta} + \mathbf{Z}\mathbf{b} + \boldsymbol{\varepsilon}$$
$$\mathbf{b} \sim \mathcal{N}(\mathbf{0}, \mathbf{D})$$
$$\boldsymbol{\varepsilon} \sim \mathcal{N}(\mathbf{0}, \mathbf{R})$$

where

- Y is the vector of height measurements. The basic unit of Y will be a measurement within a tree within a forest. It has 542 observations.
- $\mathbf{X}$ is a matrix of 0s, 1s, and diameters to allocate the observations to different tree diameters at the time of measurement.
- β is a vector of parameter estimates.
- Z is a matrix of 0s and 1s to allocate the observations to different forests and trees within forests.
- b is a vector of means for the forests and the trees.
- $\mathbf{D}$ is a block diagonal matrix that comprises two portions: a 9×9 identity matrix multiplied by a constant σ_f^2 and then a square matrix for each forest, which is a diagonal matrix with variances on the diagonals.
- $\mathbf{R}$ is a 542×542 identity matrix multiplied by a constant σ^2.

This model is fitted using the following code:

```
> hd.lme.3 <- lme(height.m ~ dbhib.cm,
+                  random = ~1 | Forest.ID/Tree.ID,
+                  data = stage)
```

The key assumptions that we're making are that:

1. The model structure is correctly specified.
2. The tree and forest random effects are normally distributed.
3. The tree random effects are homoskedastic within the forest random effects.
4. The innermost residuals are normally distributed.
5. The innermost residuals are homoskedastic within and across the tree random effects.
6. The innermost residuals are independent within the trees.
7. The samples represent the populations from which they were drawn.

We again construct diagnostic graphs to check these assumptions. To examine each of the assumptions in turn, we have constructed the earlier suite of graphics along with some supplementary graphs. The extra graphs are:

1. An extra q-q plot of the tree-level random effects to check whether they are normally distributed with constant variance.

   a. Do the points follow a straight line, or do they exhibit skewness or kurtosis? If not, then careful thought about the variance model might be beneficial.

b. Are any outliers evident? If so, they should be checked and their effect
on the model fit examined by their deletion.

2. A notched boxplot of the tree-level random effects by the grouping variable
to see what the within-group distribution looks like.

   a. Do the notches intersect 0? If not, then the relevant groups should be
checked and the effect of their deletion on the model fit examined.

   b. Is there a trend between the medians of the within-group residuals and
the estimated random effect? If so, then the random effect may be explaining some variation that could be explained by a fixed effect.

   c. Do any of the random effects stand out as being wildly different? If so,
they should be checked and the effect of their deletion on the model fit
examined.

3. A scatterplot of the variance of the tree-level random effects within the
forest against the forest random effect.

   a. Is there a distinct positive or negative trend? If so, then careful thought
about the variance model might be beneficial.

   b. Do any of the random effects stand out as being wildly different? If so,
they should be checked and the effect of their deletion on the model fit
examined.

4. An autocorrelation plot of the within-tree errors.

   a. Is there evidence of substantial autocorrelation? If so, then the model
should be extended to cover that possibility, as documented below.

As a rule of thumb, we will need about four plots plus three for each
random effect. Cross-reference these against Figures 7.7–7.9. Each graphic
should ideally be examined separately in its own frame. The code follows.

```
> opar <- par(mfrow = c(1, 3), mar = c(4, 4, 3, 1), las = 1,
+              cex.axis = 0.9)
> plot(fitted(hd.lme.3, level=0), stage$height.m,
+       xlab = "Fitted Values", ylab = "Observed Values",
+       main = "Model Structure (I)")
> abline(0, 1, col = "gray")
> scatter.smooth(fitted(hd.lme.3),
+                residuals(hd.lme.3, type="pearson"),
+                main = "Model Structure (II)",
+                xlab = "Fitted Values",
+                ylab = "Innermost Residuals")
> abline(h = 0, col = "gray")
> acf.resid <- ACF(hd.lme.3, resType = "normal")
> plot(acf.resid$lag[acf.resid$lag < 10.5],
+       acf.resid$ACF[acf.resid$lag < 10.5],
+       type="b", main="Autocorrelation",
```

```
+        xlab="Lag", ylab="Correlation")
> stdv <- qnorm(1 - 0.01/2)/sqrt(attr(acf.resid, "n.used"))
> lines(acf.resid$lag[acf.resid$lag < 10.5],
+        stdv[acf.resid$lag < 10.5],
+        col="darkgray")
> lines(acf.resid$lag[acf.resid$lag < 10.5],
+        -stdv[acf.resid$lag < 10.5],
+        col="darkgray")
> abline(0,0,col="gray")
> par(opar)
```

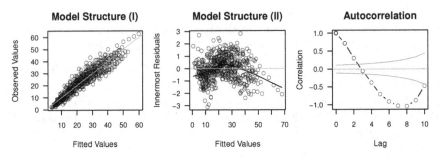

Fig. 7.7: Selected overall diagnostics for the fit of the height/diameter mixed-effects model `hd.lme.3` for Stage's data.

```
> opar <- par(mfrow = c(1, 3), mar = c(4, 4, 3, 1), las = 1,
+             cex.axis = 0.9)
> ref.forest <- ranef(hd.lme.3, level=1, standard=T)[[1]]
> ref.tree <- ranef(hd.lme.3, level=2, standard=T)[[1]]
> ref.tree.frame <- ranef(hd.lme.3, level=2,
+                         augFrame=TRUE, standard=TRUE)
> ref.var.tree <- tapply(residuals(hd.lme.3, type="pearson",
+                                  level=1),
+                         stage$Tree.ID, var)
> ref.var.forest <- tapply(ref.tree,
+                          ref.tree.frame$Forest, var)
> qqnorm(ref.forest, main = "QQ plot: Forest")
> qqline(ref.forest)
> qqnorm(ref.tree, main = "QQ plot: Tree")
> qqline(ref.tree)
> qqnorm(residuals(hd.lme.3, type="pearson"),
+        main="QQ plot: Residuals")
> qqline(residuals(hd.lme.3, type="pearson"), col="red")
> par(opar)
```

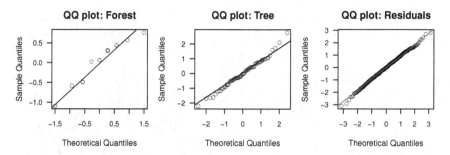

Fig. 7.8: Selected quantile-based diagnostics for the fit of the height/diameter mixed-effects model `hd.lme.3` for Stage's data.

```
> opar <- par(mfrow = c(2, 2), mar = c(4, 4, 3, 1),
+              las = 1, cex.axis = 0.9)
> boxplot(ref.tree ~ ref.tree.frame$Forest,
+         ylab = "Tree Effects", xlab = "National Forest",
+         notch= TRUE, varwidth = TRUE, at = rank(ref.forest))
> axis(3, labels=format(ref.forest, dig=2), cex.axis=0.8,
+      at=rank(ref.forest))
> abline(h=0, col="darkgreen")
> boxplot(residuals(hd.lme.3, type="pearson", level = 1) ~
+         stage$Tree.ID,
+         ylab = "Innermost Residuals", xlab = "Tree",
+         notch = TRUE, varwidth = TRUE, at=rank(ref.tree))
> axis(3, labels=format(ref.tree, dig=2), cex.axis=0.8,
+      at=rank(ref.tree))
> abline(h=0, col="darkgreen")
> plot(ref.forest, ref.var.forest, xlab="Forest Random Effect",
+      ylab="Variance of within-Forest Residuals")
> abline(lm(ref.var.forest ~ ref.forest))
> plot(ref.tree, ref.var.tree, xlab="Tree Random Effect",
+      ylab="Variance of within-Tree Residuals")
> abline(lm(ref.var.forest ~ ref.forest))
> par(opar)
```

Everything in these figures looks good except for the residual plots and the correlation of the within-tree residuals, which show an unacceptably strong signal. At this point, one might think that the next step is to try to fit an autocorrelation function to the within-tree residuals, but the kink in the residual plot suggests that it seems more valuable to take a look at a different diagnostic first.

Note that our interpretation of the diagnostics depends on what we see in other diagnostics. Also, note that much of the direction that we take is contingent upon our interpretation of the diagnostics that we examine. Providing

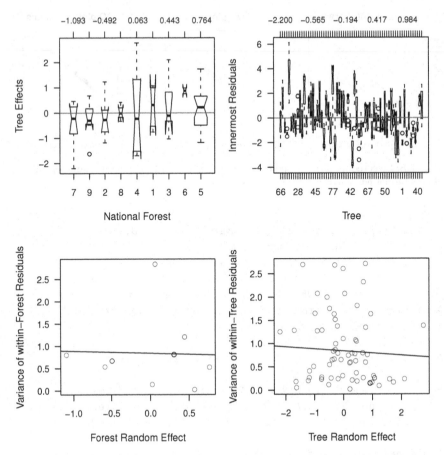

Fig. 7.9: Selected random-effects-based diagnostics for the fit of the height/diameter mixed-effects model **hd.lme.3** for Stage's data.

an unequivocal algorithm for model development, even in such a seemingly simple case, is impossible.

The augmented prediction plot overlays the fitted model with the observed data at an optional level within the model. It is constructed using xyplot from lattice package and accepts arguments that are relevant to that function for further customization. This allows us to sort the trees by national forest to help us pick up any cluster effects.

```
> library(lattice)
> trees.in.forests <-
+     with(stage, aggregate(x = list(measures = height.m),
+                           by = list(tree = Tree.ID,
+                                     forest = Forest.ID),
```

```
+                        FUN = length))
> panel.order <-
+    rank(as.numeric(as.character(trees.in.forests$tree)))

> plot(augPred(hd.lme.3),
+          index.cond = list(panel.order),
+          strip = strip.custom(par.strip.text = list(cex = 0.5)))
```

The augmented prediction plot (Figure 7.10) shows that a number of the trees have curvature in the relationship between height and diameter that the model fails to pick up, whereas others seem quite linear. It also shows that the omission of a random slope appears to create problematic lack of fit. Finally, there does not appear to be any particular pattern among the forests.

At this point, we have several options, each of which potentially leads to different resolutions of our problem or, more likely, to several further problems. How we proceed depends on our goal. We can

1. add a quadratic fixed effect;
2. add a quadratic random effect;
3. add quadratic fixed and random effects;
4. correct the model by including within-tree correlation; and
5. switch to non-linear mixed-effects models and use a more appropriate functional form.

Since we do not believe that the true relationship between height and diameter could reasonably be a straight line, we add a fixed quadratic diameter effect and a random diameter effect, by tree, and see how things go. We will show only a sample of the diagnostic graphs here.

```
> hd.lme.4 <- lme(height.m ~ dbhib.cm + I(dbhib.cm^2),
+                   random = list( ~ 1 | Forest.ID,
+                                       ~ dbhib.cm | Tree.ID),
+                   control = lmeControl(maxIter = 500,
+                                            msMaxIter = 500),
+                   data = stage)
```

Note the splitting of the random-effects statement into a list in order to permit different inclusions at different levels of the hierarchy. During the process of writing this text, we tried various more detailed arrangements, but the more complex models failed to converge for these data. Also note our invocation of the `control = lmeControl` argument in the function call. This argument is an essential element of the model-fitter's toolbox, and we delay its discussion until Section 7.4. We present the first set of relevant diagnostics in Figure 7.11.

```
> opar <- par(mfrow = c(1, 3), mar = c(4, 4, 3, 1), las = 1,
+          cex.axis = 0.9)
```

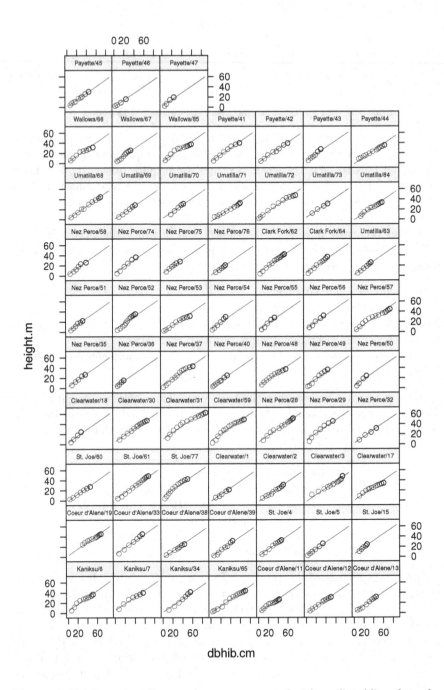

Fig. 7.10: Height against diameter by tree, augmented with predicted lines from the mixed-effects model `hd.lme.3`.

```
>   plot(fitted(hd.lme.4, level=0), stage$height.m,
+       xlab = "Fitted Values", ylab = "Observed Values",
+       main = "Model Structure (I)")
>   abline(0, 1, col = "gray")
>   scatter.smooth(fitted(hd.lme.4),
+                         residuals(hd.lme.4, type="pearson"),
+                  main = "Model Structure (II)",
+                  xlab = "Fitted Values",
+                  ylab = "Innermost Residuals")
>   abline(0, 0, col = "gray")
> acf.resid <- ACF(hd.lme.4, resType = "n")
> plot(acf.resid$lag[acf.resid$lag < 10.5],
+       acf.resid$ACF[acf.resid$lag < 10.5],
+       type="b", main="Autocorrelation",
+       xlab="Lag", ylab="Correlation")
> stdv <- qnorm(1 - 0.01/2)/sqrt(attr(acf.resid, "n.used"))
> lines(acf.resid$lag[acf.resid$lag < 10.5],
+       stdv[acf.resid$lag < 10.5],
+       col="darkgray")
> lines(acf.resid$lag[acf.resid$lag < 10.5],
+       -stdv[acf.resid$lag < 10.5],
+       col="darkgray")
> abline(0,0,col="gray")
> par(opar)
```

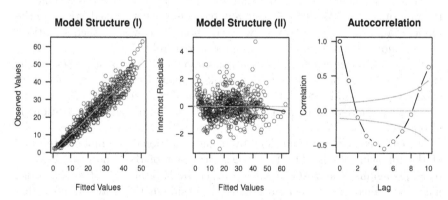

Fig. 7.11: Selected diagnostics for the fit of the height/diameter mixed-effects model
hd.lme.4 for Stage's data.

This inclusion has improved the model somewhat based on the lack of
curvature in the plot of the innermost residuals against the fitted values,
but it looks like we do need to include some accounting for the within-tree

correlation. Pinheiro and Bates (2000) detail the options that are available. Also, we'll start to use **update** because it simplifies the model expression considerably.

```
> hd.lme.5 <- update(hd.lme.4, correlation = corCAR1())
```

We now check the model using the same graphics (Figure 7.12). The necessary code is omitted to save space. The unaccounted within-tree autocorrelation is now negligible.

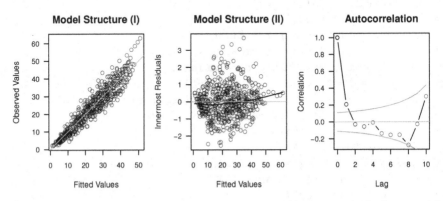

Fig. 7.12: Selected diagnostics for the fit of the height/diameter mixed-effects model `hd.lme.5` for Stage's data.

Another element of the model that we have control over is the variance of the random effects. We haven't seen any red flags for heteroskedasticity in the model diagnostics, so we haven't worried about it. However, such situations are common enough to make an example worthwhile.

Two kinds of heteroskedasticity are common and worthy of concern: first, that the variance of the response variable is related to the mean of the response variable; and second, that the conditional variance of the observations varied within one or more strata. Some combination of the two conditions is also possible.

We can detect these conditions by using conditional residual scatterplots of the following kinds. The first is a scatterplot of the innermost Pearson residuals against the fitted values, stratified by habitat type. *Pearson* residuals, also called *standardized* residuals, are residuals that have been divided by their estimated standard deviation. The code to create this graphic is part of the nlme package.

```
> plot(hd.lme.5, resid(.) ~ fitted(.) | Hab.ID, layout=c(1, 5))
```

The second is a quantile plot of the innermost Pearson residuals against the normal distribution, stratified by habitat type. This code is provided by the lattice package, and we found a template under ?qqmath (Sarkar, 2010).

```
> qqmath(~ resid(hd.lme.5) | stage$Hab.ID,
+           prepanel = prepanel.qqmathline,
+           panel = function(x, ...) {
+               panel.qqmathline(x, distribution = qnorm)
+               panel.qqmath(x, ...)
+           })
```

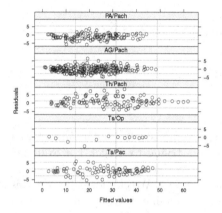 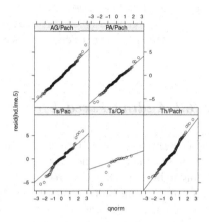

Fig. 7.13: Innermost Pearson residuals against fitted values by habitat type.

Fig. 7.14: Quantile plots of innermost Pearson residuals against the normal distribution by habitat type.

There seems little evidence in either Figure 7.13 or Figure 7.14 to suggest that the variance model is problematically incorrect. The TsOp panel of the q-q plot looks a little peculiar, but the sample size is very small. We are not perturbed. Had the variance model seemed inadequate, we could have used the weights argument in a call to update with one of the following approaches:

- weights = varIdent(form=~1 | HabType.ID). This option allows the observations within each habitat type to have their own variance.
- weights = varPower(). This option fits a power function for the relationship between the variance and the predicted means, and estimatse the exponent.
- weights = varPower(form = ~dbhib.cm | HabType.ID). This option fits a power function for the relationship between the variance and the diameter uniquely within each habitat type and estimates the exponent.
- weights = varConstPower() This option fits a power function with a constant for the relationship between the variance and the predicted mean and estimates the exponent and constant.

Other options are available; the function is fully documented in Pinheiro and Bates (2000). We accept the model as it stands for the moment. This model is the baseline model, as it provides predictions of height from diameter and satisfies the regression assumptions. Other models may later prove to be better fitting; for example it may be that including habitat type or age in the model obviates our use of the quadratic diameter term. Whether or not this makes for a better model in terms of actual applications will vary!

We now use the following code to construct a snapshot of the fitted lines and the observations (Figure 7.15).

```
> plot(augPred(hd.lme.5),
+        index.cond = list(panel.order),
+        strip = strip.custom(par.strip.text = list(cex = 0.5)))
```

7.3.3 Adding Fixed Effects

We can try to extend the baseline model to improve its performance based on our knowledge of the system that we are trying to model. For example, it might be true that the tree age mediates its diameter–height relationship in a way that has not been captured in the model. We can formally test this assertion using the **anova** function, we can examine it graphically using an added-variable plot, or we can try to fit the model with the term included and assess what effect the extra term has on the residual variation.

An added-variable plot is a graphical summary of the amount of variation that is uniquely explained by a predictor variable. It can be constructed in R as follows. Here we need to decide what level of residuals to choose, as there are several. We adopt the outermost residuals.

```
> age.lme.1 <- lme(Age ~ dbhib.cm,
+                   random = ~1 | Forest.ID/Tree.ID,
+                   data = stage)
> res.Age <- residuals(age.lme.1, level = 0)
> res.HD <- residuals(hd.lme.5, level = 0)
> scatter.smooth(res.Age, res.HD,
+     xlab = "Variation unique to Age",
+     ylab = "Variation in Height after all but Age")
```

In order to assess whether we would be better served by adding habitat type to the model, we can construct a graphical summary of observed against predicted heights arranged by habitat type (Figure 7.16)

```
> xyplot(stage$height.m ~ fitted(hd.lme.5, level=0) | Hab.ID,
+         xlab="Predicted height (m)",
+         ylab="Observed height (m)",
+         data=stage,
+         panel = function(x, y, subscripts) {
```

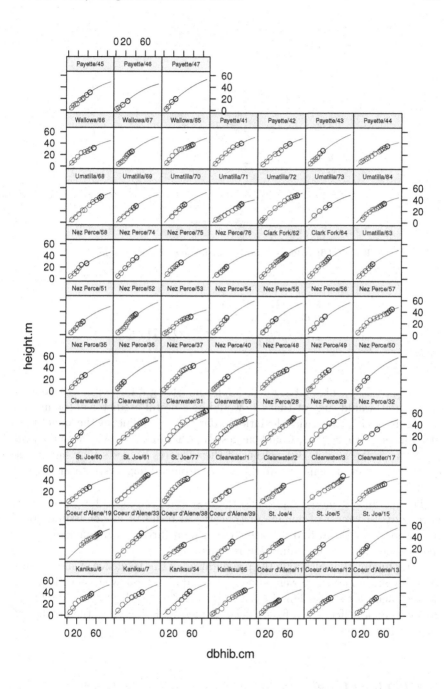

Fig. 7.15: Height against diameter by tree, augmented with predicted lines.

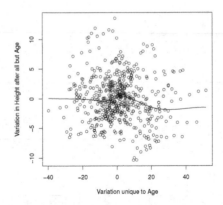

Fig. 7.16: Added-variable plot for age against height=.

```
+                       panel.xyplot(x, y)
+                       panel.abline(0, 1)
+                       panel.abline(lm(y ~ x), lty=3)
+            }
+ )
```

In Figure 7.16, substantial deviation from a straight line with zero slope, if it occurred, would suggest that age would be a useful predictor in the model. In Figure 7.17, substantial deviation from the $y = x$ line in the panel-wise fitted lines, if it occurred, would suggest that habitat type would be a useful predictor in the model. But neither of these figures suggests that significant or important improvements would accrue from adding these terms to the model.

The incumbent model represents the best compromise so far. It seems to have addressed most of our major concerns in terms of model assumptions. It may be possible to find a better model with further searching. However, there comes a point of diminishing returns.

Note, finally, that although the presentation of this sequence of steps seems fairly linear, in fact there were numerous blind alleys followed, much looping, and retracing of steps. This is neither a quick nor a direct process! Introducing random effects to a fixed-effects model vastly increases the number of diagnostics to check and possibilities to follow.

7.3.4 The Model

Let us now examine our final model. We have fit

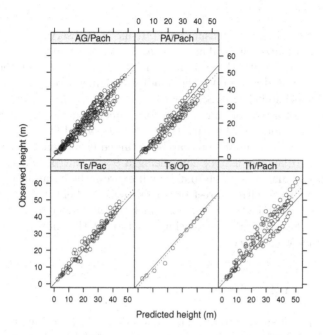

Fig. 7.17: Plot of predicted height against observed height by habitat type. The solid line is 1:1, as predicted by the model. The dotted line is the OLS line of best fit within habitat type.

$$y_{ijk} = \beta_0 + b_{0i} + b_{0ij}$$
$$+ \left(\beta_1 + b_{1ij}\right) \times x_{ijk}$$
$$+ \beta_2 \times x_{ijk}^2$$
$$+ \varepsilon_{ijk}$$

where y_{ijk} is the height measured at time k for the j-th tree in the i-th forest and x_{ijk} is the diameter measured at time k for the j-th tree in the i-th forest; β_0, β_1, and β_2 are fixed population-level parameters; b_{0i} are forest-specific intercepts; b_{0ij} and b_{1ij} are tree-specific intercepts and slopes, respectively; and ε_{ijk} are time- and tree-specific errors. In matrix form, the model expression is still

$$\mathbf{Y} = \mathbf{X}\boldsymbol{\beta} + \mathbf{Z}\mathbf{b} + \boldsymbol{\varepsilon}$$
$$\mathbf{b} \sim \mathcal{N}\left(\mathbf{0}, \mathbf{D}\right)$$
$$\boldsymbol{\varepsilon} \sim \mathcal{N}\left(\mathbf{0}, \mathbf{R}\right)$$

The structure is as follows:

- Y is the vector of height measurements. It has 542 observations.
- X is a 542×3 matrix of 1s, diameters, and squared diameters.
- β is a vector of length three: it has an intercept, a slope for the linear diameter term, and a slope for the quadratic diameter term.
- Z is a 542×141 unit brute. See below.
- b is a vector of intercepts and slopes for diameter and diameter squared for each forest, and then for each tree. It will be $9 + 132 = 141$ elements long. See below. The predictions can be obtained using `ranef(hd.lme.5)`.
- D is a block diagonal matrix comprising a 9×9 matrix followed by 66 2×2 identical matrices. Each matrix will express the covariance between the random effects within forest or within tree. See below.
- R is a 542×542 symmetric matrix for which the off diagonals are 0 between trees and a geometrically decreasing function of the inter-measurement time distance within trees.

7.3.4.1 More on Z

Recall that the role of Z is to allocate the random effects to the appropriate element. This allocation can be somewhat complicated. Our Z can be divided into two distinct sections: a 542×9 matrix Z_f associated with the forest-level effects and a 542×132 matrix Z_t associated with the tree-level effects. In matrix nomenclature,

$$Z = [Z_f \mid Z_t] \tag{7.11}$$

Now, Z_f allocates the random intercept to each observation from each forest. There are nine forests, so any given row of Z_f will contain eight zeros and a 1. Similarly, Z_t allocates an intercept and slope to each observation from each tree. There are 66 trees, so any given row of Z_t will contain 195 zeros, a 1, and the corresponding *dbhib*.

7.3.4.2 More on b

The purpose of b is to contain all the predicted random effects. Thus it will be 141 units long, which corresponds to 1 unit for each level of forest (intercept) and 2 units for each level of tree (intercept, slope for diameter).

$$b = (b_{f1}, b_{f20}, \ldots, b_{t10}, b_{t1d}, b_{t20}, b_{t2d}, \ldots)' \tag{7.12}$$

The combination of b and Z serves to allocate each random effect to the appropriate unit and measurement.

7.3.4.3 More on D

Finally, **D** dictates the relationships between the different random effects within the levels of forest and tree. We have assumed that the random effects will be independent between habitat types and trees. So, there are only two submatrices to this matrix, called $\mathbf{D_f}$ and $\mathbf{D_t}$.

$$\mathbf{D_f} = \left[\sigma_{bf0}^2 \right] \tag{7.13}$$

$$\mathbf{D_t} = \begin{bmatrix} \sigma_{bt0}^2 & \sigma_{bt0d} \\ \sigma_{bt0d} & \sigma_{btd}^2 \end{bmatrix} \tag{7.14}$$

Then the structure of **D** is simply nine repetitions of $\mathbf{D_f}$ laid on a diagonal line followed by 66 repetitions of $\mathbf{D_t}$ laid on the same diagonal and zeros everywhere else.

7.4 Model Wrangling

We observed earlier that the use of the `control` argument was a key tool for the modeler. This element can introduce a little culture shock. Having come from traditions of model fitting for which exact solutions were easily obtained and convergence was unequivocal, it was surprising, not to say disheartening, for us to find that algorithms sometimes quit before they achieved convergence. Probably we are displaying our naivete.

The statistical models that we have been discussing in this chapter do not have exact solutions. Accordingly, we have to try to maximize the likelihood, for example, by iterative means. It is necessary and correct that the authors of the code we use will have put in checks to halt the code in situations where they deem continuing to be unprofitable.

In any case, bitter experience and ruthless experimentation have taught us that the code authors do not necessarily have exactly our problem in mind when they are choosing the default parameters for their software. In such cases, it is necessary to roll up our sleeves and plunge our arms into the organs of our analysis. Most of the fitting tools that we use have control arguments that will report or modify the process of model fitting. Experimenting with them will often lead to model configurations that fit reliably.

In short, don't be reluctant to experiment. Any or all of the following strategies might be necessary to achieve a satisfactory fit of your model to your data.

7.4.1 Monitor

In order to be better informed about the progress of model fitting, we use the
`msVerbose` argument. It provides a brief updating description of the progress
of the model fit. It will also point out problems along the way, which help
the user decide what is the best thing to do next.

7.4.2 Meddle

This strategy involves adjusting the fitting tool.

If the model is failing to converge, then often all that is required is an
increase in the number of allowable iterations. The mixed-effects model fitting
algorithm in `lme` uses a hybrid optimization scheme that starts with the EM
algorithm and then changes to Newton–Raphson (Pinheiro and Bates, 2000,
p. 80). The latter algorithm is implemented with two loops, so we have three
iteration caps. We have found that increasing both `maxIter` and `msMaxIter`
is a useful strategy. If we are feeling patient, we will increase them to about
10000 and monitor the process to see if the algorithm still wishes to search.
We have occasionally seen iteration counts in excess of 8000 for models that
subsequently converged.

We have also had success with changing the optimization algorithm. That
is, models that have failed to converge with `nlminb`, by getting caught in a
singular non-convergence, have converged successfully using `Nelder-Mead` in
`optim`. The default is to use `nlminb`, but it may be worth switching to `optim`,
and within `optim` choosing between `Nelder-Mead` and `BFGS`. Each of these
algorithms has different properties and different strengths and weaknesses.
Any might lead more reliably to a satisfactory solution.

7.4.3 Modify

This strategy involves changing the relationship between the model and the
data.

A number of the model components permit the specification of a starting
point. For example, if we provide the `corAR1` function with a suitable number, then the algorithm will use that number as a starting point. Specifying
this value can help the algorithm converge speedily or at all. Experimenting
with subsets of the full model to try to find suitable starting points can be
profitable; for example, if one has a correlation model and a variance model.

We can also think about how the elements in the data might be interacting
with the model. Is the dataset unbalanced or does it contain outliers, or is it
too small? Any of these conditions can cause problems for fitting algorithms.

Examining the data before fitting any model is standard practice. Be prepared to *temporarily* delete data points, or *temporarily* augment underrepresented portions, in order to provide a reasonable set of starting values to those functions that accept them.

7.4.4 Compromise

Sometimes a model involves a complicated hierarchy of random effects. It is worth asking whether or not such depth is warranted and whether a superficially more complex but simpler model might suffice. The case study in this chapter serves as a good example: although model fit benefited by allowing each individual tree to have a random slope, there was no need to allow each national forest to have a random slope. Including a slope for each forest made the model unnecessarily complicated and also made fitting the model much harder. Specifying the smaller model was a little less elegant, however.

Finally, sometimes no matter what exigencies we try, a model will not converge. There is a point in every analysis where we must decide to cut our losses and go with the model we have. If we know that the model has shortcomings, then it is our responsibility to draw attention to those shortcomings. For example, if we are convinced that there is serial autocorrelation in our residuals but cannot achieve a reasonable fit using the available resources, then providing a diagnostic plot of that autocorrelation is essential. Furthermore, it is important to comment on the likely effect of the model shortcoming upon inference and prediction. If we are fortunate enough to be able to fit a simpler model that does include autocorrelation, for example, we might demonstrate what effect the inclusion of that portion of the model has upon our conclusions. We would do this by fitting three models: the complex model, the simple model with the autocorrelation, and the simple model without the autocorrelation. If the difference between the latter two models is modest, then we have some modest degree of indirect evidence that perhaps our conclusions will be robust to misspecification of the complex model. It is not ideal, but we must be pragmatic.

7.5 The Deep End

We will now provide some mathematical motivation for the fitting strategies that we showcased in the previous section. There are numerous different representations of the linear mixed-effects model. We adopt that suggested by Laird and Ware (1982):

$$\mathbf{Y} = \mathbf{X}\boldsymbol{\beta} + \mathbf{Z}\mathbf{b} + \boldsymbol{\varepsilon}$$
$$\mathbf{b} \sim \mathcal{N}(\mathbf{0}, \mathbf{D})$$
$$\boldsymbol{\varepsilon} \sim \mathcal{N}(\mathbf{0}, \mathbf{R})$$

Here $\mathbf{Y}$ and $\mathbf{X}$ are the response variable and the design matrix, respectively, and $\mathbf{D}$ and $\mathbf{R}$ are variance–covariance matrices that are preferably constructed using a small number of parameters, which will be estimated from the data. That is, despite their apparent complexity, they are motivated by a handful of parameters, such as σ_{b01}^2, σ_{b02}^2, and $\sigma_{b_1}^2$, as above.

7.5.1 Maximum Likelihood

Recall that the principle behind maximum likelihood was to find the set of parameter estimates that were best supported by the data. This began by writing down the conditional distribution of the observations. For example, the pdf for a single observation from the normal distribution is

$$f\left(y_i \mid \mu, \sigma^2\right) = \frac{1}{\sqrt{2\pi}\sigma} e^{\frac{-(y_i - \mu)^2}{2\sigma^2}}$$

So a vector of observations Y is distributed according to $Y \overset{\mathrm{d}}{=} \mathrm{N}(\mu, \mathbf{V})$, and

$$f(\mathbf{Y} \mid \mu, \mathbf{V}) = \frac{|\mathbf{V}|^{-\frac{1}{2}}}{(2\pi)^{\frac{n}{2}}} e^{-\frac{1}{2}(\mathbf{Y}-\mu)'\mathbf{V}^{-1}(\mathbf{Y}-\mu)}$$

Now if we are interested in predicting Y using some linear predictors X, we might claim that the expectation of Y is equal to some linear combination of the Xs. So in terms of the linear model $\mathbf{Y} = \mathbf{X}\boldsymbol{\beta}$, the conditional joint density is

$$f(\mathbf{Y} \mid \mathbf{X}, \boldsymbol{\beta}, \mathbf{V}) = \frac{|\mathbf{V}|^{-\frac{1}{2}}}{(2\pi)^{\frac{n}{2}}} e^{-\frac{1}{2}(\mathbf{Y}-\mathbf{X}\boldsymbol{\beta})'\mathbf{V}^{-1}(\mathbf{Y}-\mathbf{X}\boldsymbol{\beta})}$$

Reversing the conditioning and taking logarithms yields the log-likelihood:

$$\mathscr{L}(\boldsymbol{\beta}, \mathbf{V} \mid \mathbf{Y}, \mathbf{X}) = -\frac{1}{2}\ln(|\mathbf{V}|) - \frac{n}{2}\ln(2\pi) - \frac{1}{2}(\mathbf{Y} - \mathbf{X}\boldsymbol{\beta})'\mathbf{V}^{-1}(\mathbf{Y} - \mathbf{X}\boldsymbol{\beta})$$

Notice that the parameters we are interested in, $\boldsymbol{\beta}$ and $\mathbf{V}$, are now embedded in the likelihood. Solving for those parameters requires maximizing the likelihood. Assume that we know $\mathbf{V}$. Then, to find $\hat{\boldsymbol{\beta}}$, we take the derivative of $\mathscr{L}(\boldsymbol{\beta}, \mathbf{V} \mid \mathbf{y}, \mathbf{X})$ with regard to $\boldsymbol{\beta}$:

$$\frac{\mathrm{d}\mathscr{L}}{\mathrm{d}\beta} = \frac{\mathrm{d}}{\mathrm{d}\beta}\left[-\frac{1}{2}\left(\mathbf{y}-\mathbf{X}\beta\right)'\mathbf{V}^{-1}\left(\mathbf{y}-\mathbf{X}\beta\right)\right]$$

This formulation leads, as we've seen earlier, to

$$\hat{\beta}_{MLE} = \left(\mathbf{X}'\mathbf{V}^{-1}\mathbf{X}\right)^{-1}\mathbf{X}'\mathbf{V}^{-1}\mathbf{Y}$$

but this can only be solved *if we know* **V**! Since we do not know **V**, we have to maximize the likelihood as follows. First, substitute

$$\left(\mathbf{X}'\mathbf{V}^{-1}\mathbf{X}\right)^{-1}\mathbf{X}'\mathbf{V}^{-1}\mathbf{Y}$$

for β in the likelihood. That is, remove all the instances of β, and replace them with this statement. By this means, β is *profiled out* of the likelihood. The likelihood is now only a function of the data and the covariance matrix V. This covariance matrix is itself a function of the covariance matrices of the random effects, which are structures that involve hopefully only a few unknown parameters and are organized by the model assumptions.

We maximize the resulting log-likelihood in order to estimate $\hat{V}$, and then we calculate the estimates of the fixed effects via

$$\hat{\beta}_{MLE} = \left(\mathbf{X}'\hat{\mathbf{V}}^{-1}\mathbf{X}\right)^{-1}\mathbf{X}'\hat{\mathbf{V}}^{-1}\mathbf{Y} \tag{7.15}$$

After some algebra, which is well documented in, for example, Schabenberger and Pierce (2002), we also get the best linear unbiased predictors (BLUPs) of the random effects,

$$\hat{b}_{MLE} = \mathbf{D}\mathbf{Z}'\hat{\mathbf{V}}\left(\mathbf{Y}-\mathbf{X}\hat{\beta}\right) \tag{7.16}$$

where **D** is the covariance matrix of the random effects.

7.5.2 Restricted Maximum Likelihood

Maximum-likelihood estimators of covariance parameters are usually negatively biased. *Restricted* or *residual* maximum likelihood will penalize the variance estimates based on the model size and is therefore preferred for mixed-effects models. REML-based estimates are not unbiased, except under certain circumstances, but they are expected to be less biased than maximum-likelihood estimates. See Demidenko (2004) for a useful discussion.

Instead of maximizing the conditional joint likelihood of **Y**, we do so for an (almost) arbitrary linear transformation of **Y**, which we shall denote **K**. It is almost arbitrary inasmuch as there are only two constraints: **K** must have full column rank, or else we would be creating observations out of thin air, and **K** must be chosen so that $E[\mathbf{K}'\mathbf{Y}] = 0$.

The easiest way to guarantee that these hold is to ensure that $\mathbf{K}'\mathbf{X} = 0$ and that $\mathbf{K}$ has no more than $n - p$ independent columns, where p is the number of independent parameters in the model. This removes the fixed effects from consideration and in so doing also penalizes the estimation for model size. So, the likelihood is restricted by the fixed effects being set to 0, thus the name *restricted* maximum likelihood.

So, briefly, REML involves applying ML but replacing $\mathbf{Y}$ with $\mathbf{KY}$, $\mathbf{X}$ with $\mathbf{0}$, $\mathbf{Z}$ with $\mathbf{K}'\mathbf{Z}$, and $\mathbf{V}$ with $\mathbf{K}'\mathbf{VK}$. Pawitan (2001) and Lee et al. (2006) show that the REML estimation procedure can be derived as maximization of a modified profiled likelihood.

7.6 Non-linear Mixed-Effects Models

In Section 6.2.3, we fit a nice non-linear model to a single tree. It would be good to be able to fit the same model to a collection of trees with minimal fuss. The nlme package again provides us with a way forward. First, we can use the `groupedData` structure to simplify the problem of fitting a model to many trees, and secondly, we can use a so-called self-starting function, which provides its own starting values for any data that are presented to it. This self-starting function is one of the packaged functions mentioned earlier, and its adoption simplifies our approach considerably.

Our data are Norway spruce measurements drawn from von Guttenberg (1915), kindly provided to us by Professor Boris Zeide. Our goal in the first instance is to construct a model that predicts tree diameter as a function of age. The data import and cleaning are documented in Section 2.4.4.

The lattice package provides us with an easy way to plot the data (Figure 7.18). Note that these are not particularly large trees!

```
> library(lattice)

> xyplot(dbh.cm ~ age.bh | tree.ID, type="l", data=gutten)
```

```
> library(nlme)
> gutten.d <- groupedData(dbh.cm ~ age.bh | tree.ID,
+                         data = gutten)
```

The relevant self-starting function is called `SSasympOrig`.

```
> gutten.nlsList <-
+   nlsList(dbh.cm ~ SSasympOrig(age.bh, asymptote, scale),
+          data = gutten.d)
```

We are now in a position to speculate as to whether or not the kink that we observed in the residuals for tree 1.1 is repeated in the other trees (Figure 7.19). Figure 7.19 is created using the following code.

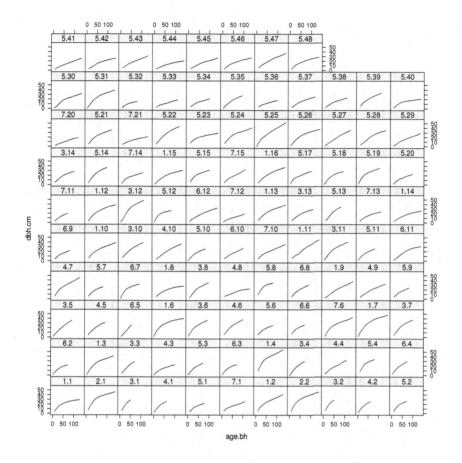

Fig. 7.18: Tree diameter data from von Guttenberg.

```
> plot(gutten.nlsList,
+        residuals(., type="pearson") ~ fitted(.) | tree.ID)
```

These results suggest that there certainly is a systematic lack of fit across the board. We may need to adopt a more flexible function.

We could print out all the parameter estimates but it is more useful to construct a graphical summary, using the following commands (Figure 7.20).

```
> plot(intervals(gutten.nlsList), layout=c(2,1))
```

Note that the intervals are based on large-sample theory.

We can extract and manipulate the coefficient estimates with a little bit of digging. The digging follows, and may be skipped. First we try to find what functions are available to manipulate objects that have the same class as the object that we have just created.

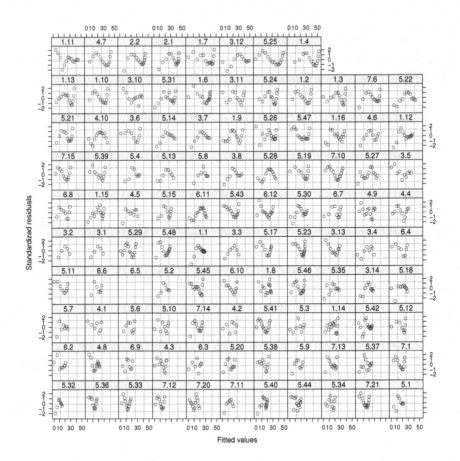

Fig. 7.19: Plot of residuals against fitted values from non-linear models as fit to each tree.

```
> methods(class=class(gutten.nlsList))
```

```
[1] formula.nlsList* nlme.nlsList      summary.nlsList*
[4] update.nlsList*
```

```
    Non-visible functions are asterisked
```

A summary method is available. Let's find out what functions are available to manipulate summary objects.

```
> methods(class=class(summary(gutten.nlsList)))
```

```
[1] coef.summary.nlsList*
```

```
    Non-visible functions are asterisked
```

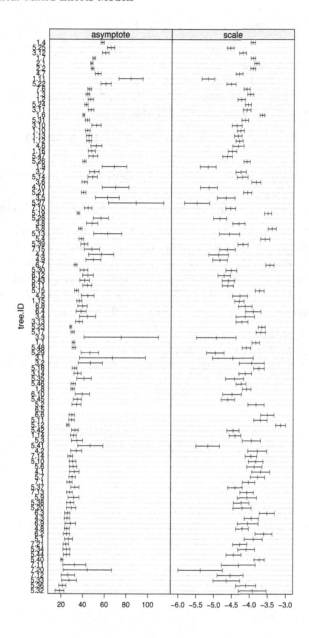

Fig. 7.20: Interval plot for diameter prediction model fitted to the von Guttenberg data.

A `coef` method is available. What does its output look like?

```
> str(coef(summary(gutten.nlsList)))
```

```
num [1:107, 1:4, 1:2] 18.7 21.4 27.6 26.5 44.5 ...
 - attr(*, "dimnames")=List of 3
  ..$ : chr [1:107] "5.32" "5.36" "5.33" "7.12" ...
  ..$ : chr [1:4] "Estimate" "Std. Error" "t value" "Pr(>|t|)"
  ..$ : chr [1:2] "asymptote" "scale"
```

It is an array. We can extract its elements in the following way:

```
> asymptote <-
+    coef(summary(gutten.nlsList))[,"Estimate","asymptote"]
> half.age <- log(2) /
+    exp(coef(summary(gutten.nlsList))[,"Estimate","scale"])
```

These two objects contain the parameter estimates for each tree. Figure 7.21 shows a scatterplot of the estimated parameters for each tree: the estimated asymptote on the y-axis and the estimated age at which the tree reaches half its maximum diameter on the x-axis, with a *lowess* smooth added to describe the mean of the pattern. There is clearly a relationship between the asymptote and the estimated age at which half the asymptote is reached.

```
> opar <- par(las=1, mar=c(4,4,1,1))
> scatter.smooth(half.age, asymptote,
+                xlab = "Age", ylab = "Asymptote")
> par(opar)
```

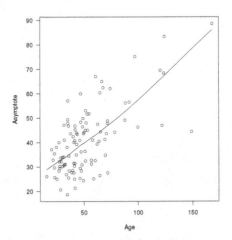

Fig. 7.21: Plot of the estimated age at which the tree reaches half its maximum diameter against the estimated tree-level maximum diameter.

This approach provides us with a convenient way to think about fitting data to many different objects. But, what practical use does the model have?

Not much. This model is analogous to the linear model that includes one intercept for each plot: the model assumptions are (probably) satisfied, but the model isn't really useful. We need a way to tie all these little models together. Hence, we adopt a hierarchical approach.

7.6.1 Hierarchical Approach

We now focus on fitting the same kinds of models to hierarchical data. This direction produces simplifications and complications, but sadly more of the latter than the former.

We use the non-linear mixed-effects model that generalizes the Pinheiro and Bates (2000) parameterization of our earlier non-linear model (equation (6.6)). We will start with allowing each tree to have a random asymptote and scale. That is, for diameter measure t in tree i,

$$y_{it} = (\phi_1 + \phi_{i1}) \times [1 - \exp(-\exp((\phi_2 + \phi_{i2})x))] + \varepsilon_{it} \qquad (7.17)$$

where ϕ_1 is the fixed, unknown asymptote and ϕ_2 is the fixed, unknown scale, and

$$\begin{bmatrix} \phi_{i1} \\ \phi_{i2} \end{bmatrix} \sim \mathcal{N}\left(\begin{bmatrix} 0 \\ 0 \end{bmatrix}, \begin{bmatrix} \sigma_1^2 & \sigma_{12} \\ \sigma_{12} & \sigma_2^2 \end{bmatrix} \right) \qquad (7.18)$$

The process of fitting and critiquing these models is well documented in Pinheiro and Bates (2000). Given that we have already used `nlsList`, the easiest approach both from the point of view of typing and of having sensible starting points is to use `gutten.nlsList` as the starting point for the model:

```
> gutten.nlme.0 <- nlme(gutten.nlsList)
```

If we wanted to construct this model from scratch, then we would need to do this:

```
> gutten.nlme.0 <-
+   nlme(dbh.cm ~ SSasympOrig(age.bh, asymptote, scale),
+       fixed = asymptote + scale ~ 1,
+       random = asymptote + scale ~ 1,
+       start = c(asymptote = 50, scale = -5),
+       data = gutten.d)
```

Note that as of the current version[2] we do need to include the estimated starting values.

As with the linear mixed-effects models, we have a wide array of different diagnostic plots that we can deploy, and the functions to obtain those

---

[2] nlme 3.1-96 on R 2.11.1.

diagnostics are pretty much identical. Here we will focus on examining the within-tree autocorrelation (Figure 7.22).

```
> plot(ACF(gutten.nlme.0, form = ~1|tree.ID), alpha=0.01)
```

The figure shows that there is substantial within-tree autocorrelation, which suggests systematic lack of fit. At this point, we have two options: we can try a different mean function or we can try to model the autocorrelation. That is, we can try to improve the fixed effects or we can try to compensate using the random effects. In general, we should try them in that order. For the moment, though, we will focus on modeling the autocorrelation. After some experimentation, we arrived at

```
> gutten.nlme.1 <- update(gutten.nlme.0,
+                         correlation = corARMA(p = 1, q = 2))
```

This model produces residuals with the autocorrelation pattern presented in Figure 7.23.

```
> plot(ACF(gutten.nlme.1, resType = "n", form = ~1|tree.ID),
+        alpha = 0.01)
```

This is clearly superior to the preceding model, but still needs some tinkering. Perhaps there is too much lack of fit?

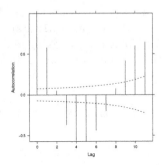

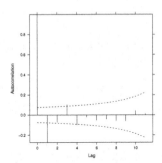

Fig. 7.22: Autocorrelation of within-tree residuals from non-linear mixed-effects model.

Fig. 7.23: Autocorrelation of within-tree residuals from non-linear mixed-effects model with explicit autocorrelation model.

As noted above, we can also try to eliminate the patterns in the residuals using a more flexible model. In the previous chapter, we tried a model that does not constrain the fitted line to pass through the origin, SSasymp. We can try this model here using the following code:

```
> gutten.nlme.2 <-
+   nlme(dbh.cm ~ SSasymp(age.bh, asymptote, scale, R0),
+        fixed = asymptote + scale + R0 ~ 1,
+        random = asymptote + scale  + R0 ~ 1,
+        start = c(asymptote = 40, scale = -0.07, R0 = -4),
+        data = gutten.d)
```

We had to try numerous different starting points before this model would converge. For example,

1. with c(asymptote=50, scale=-5, R0=1), the error was "system is computationally singular",
2. with c(asymptote=50, scale=-5, R0=-10), the error was "step halving factor reduced below minimum in PNLS step",
3. with c(asymptote=50, scale=-5, R0=-1), the model converged but took several minutes, and
4. with c(asymptote=40, scale=-0.07, R0=-4), the model converged within a few seconds.

We can compare the models using whole-model tests in the anova function. The test statistics should be regarded as only approximate. Here we see some evidence to suggest that the constraint leads to a worse-fitting model.

```
> anova(gutten.nlme.0, gutten.nlme.2)
```

	Model	df	AIC	BIC	logLik	Test
gutten.nlme.0	1	6	3627.763	3658.304	-1807.882	
gutten.nlme.2	2	10	3473.265	3524.166	-1726.632	1 vs 2

	L.Ratio	p-value
gutten.nlme.0		
gutten.nlme.2	162.4983	<.0001

A graphical comparison provides a useful addition. Because we constructed these models from a groupedData object, we can use the augmented prediction plots as a very useful summary of the model's behavior. This graphic is presented in Figure 7.24, and shows that the differences between the models does not appear to be substantial.

```
> plot(comparePred(gutten.nlme.0, gutten.nlme.2))
```

We can get the approximate 95% confidence intervals of the estimates from the following call:

```
> intervals(gutten.nlme.1)
```

```
Approximate 95% confidence intervals
```

```
 Fixed effects:
              lower        est.        upper
```

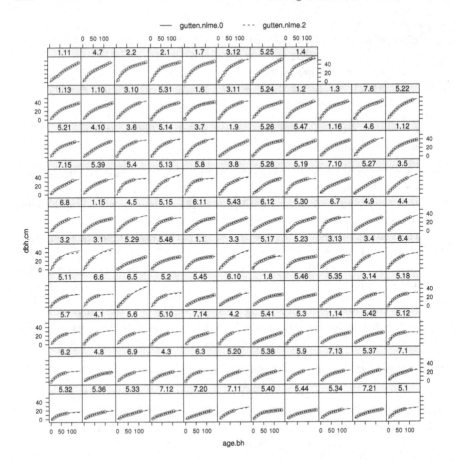

Fig. 7.24: Diameter data from von Guttenberg with both origin-constrained and unconstrained asymptotic models fitted. Here the tree-level models (equation (7.17)) are plotted.

```
asymptote 36.477392 38.649203 40.821014
scale     -4.203253 -4.119008 -4.034763
attr(,"label")
[1] "Fixed effects:"

Random Effects:
  Level: tree.ID
                          lower       est.       upper
sd(asymptote)         8.9042315 10.5142706 12.4154325
sd(scale)             0.3232857  0.3855331  0.4597661
cor(asymptote,scale) -0.6726260 -0.5225912 -0.3312863
```

```
Correlation structure:
           lower       est.      upper
Phi1    0.7549473 0.8217214 0.8716312
Theta1 0.2940804 0.4007537 0.5090448
Theta2 0.5607426 0.6827465 0.7757331
attr(,"label")
[1] "Correlation structure:"

 Within-group standard error:
   lower     est.     upper
1.291170 1.524274 1.799463
```

7.7 Further Reading

The book that documents the fitting tools that we have used for this chapter, Pinheiro and Bates (2000), should be considered essential reading for anyone wanting to use the code. The authors describe the theory, the algorithms, and most of the options, and provide substantial examples for using the code. We have also found Schabenberger and Pierce (2002), Demidenko (2004), Fitzmaurice et al. (2004), Wood (2006), Lee et al. (2006), and Gelman and Hill (2007) to be very useful. Venables and Ripley (2002) also provides some advice and examples.

Part IV
Simulation and Optimization

Chapter 8
Simulations

In this chapter, we use R to generate, examine, detect, and illuminate simulated yield tables, projections, or sets of possible future forest conditions. Our objectives are to 1) cover the major topics and tasks required to generate forest forecasts; 2) compare some common metrics from the resulting simulations; and 3) examine and present potential shortcomings and remedies for the methods presented. Our motivation is to generate simulations and combinations of simulations that can be 1) examined quickly for anomalies; 2) easily queried to answer specific questions; and 3) efficiently exported into other applications like harvest scheduling and transportation applications (Weintraub and Navon, 1976) or ecological community analysis (Oksanen et al., 2010), or linked to a geospatial database (Prayaga et al., 2009).

We tackle a specific and common problem in forest growth simulation in which more than one growth and dynamics model is needed in order to provide forecasts but those models are not commensurate, meaning that the physical scale of the unit of simulation between the models differs. Here we use the individual-tree young-stand simulator rconifers (Ritchie and Hamann, 2006, 2008) and construct a stand-level, established-stand simulator using the equations presented by Chambers (1980). The jurisdiction of the young-stand simulator presented here is normally for plantations below 25 years of age. The jurisdiction for the established-stand simulator is considered to be no younger than 30 years. This gap, or in some cases overlap, is not unusual, and it has been expected that a smooth transition between the young-stand simulator and the established-stand simulator would occur at about the time of crown closure. Here we will focus on the challenges that are introduced by our attempts to use projections from these two models jointly and ignore the minutiae of the individual jurisdictions of the respective models.

The chapter is structured as follows. In Section 8.1, we provide a brief presentation of rconifers and construct the established-stand simulator using Chambers (1980). In Section 8.2, we then determine volume and volume distributions using the taper function developed by Kozak et al. (1968) and individual tree records. In Section 8.3, we merge the results from our two models into a single data frame object. We use that object to examine the following questions in Section 8.4: 1) What, if anything, happens to the predicted maximum mean annual increment when we merge the results from multiple models (e.g., linear combination of the results)? 2) What, if anything, happens to the distribution of log product volumes when the results are merged to create a single prediction? Finally, in Section 8.5, we export the predicted yields to a database using the Structured Query Language (SQL) to allow our results to be used conveniently in other software.

8.1 Generating Simulations

The simulation of forest growth and dynamics is a complicated and multifacted undertaking. Depending on the objectives, forest simulation may require multiple software tools (e.g., SAS, PostgreSQL, Python), several simulators and models (e.g., CONIFERS, ORGANON, ZELIG), and sometimes a handful of supporting model components (e.g., taper functions, wildlife habitat models, and fuel loading models). Coordinating and linking these disparate tools is a substantial challenge.

Few forest simulation models have been developed within R, although there are some tools for simulating ecological phenomena (see, e.g., Oksanen et al., 2010). We must construct our own tools to achieve our objectives. Here we use two different and non-commensurate models that have different model architectures: 1) rconifers, a young-stand and single-plant model developed by Hamann and Ritchie (2009); and 2) the normal yield tables developed by Chambers (1980) for stand-level projection of established stands.

We start by generating model forecasts given 1) some initial set of conditions; 2) a set of possible management decisions (e.g., harvest, thin, and do nothing); and 3) a list of the required outputs (e.g., reports, maps, and files). The process, presented in Algorithm 1, comprises the steps that are used to generate the required outputs: age, volume, and volume distribution by log grade. For this simple example, the process is: 1) project the initial plant list one year, then 2) compute the volume and volume distribution by log grade, then 3) compute the summary statistics for all necessary metrics of interest, and finally 4) repeat the process until the end of the planning horizon. These procedures are the major focus of this section. In Sections 8.2 and 8.3, we examine and address potential problems.

The final step is to plot graphs, generate maps, and write reports. We assume that each project will have a unique set of metrics, constraints, and objectives, so here we focus on only three outputs for simplicity: age, volume, and volume distribution by log grade. For more complicated simulation projects, the list of inputs, management decisions, and outputs can be extensive.

The steps in Algorithm 1 present no specific challenge as individual tasks. As a complete process, however, the GrowOneYear task presents a common dilemma: How do we continue projecting the sample under different conditions? Specifically, in our project, how do we manage the transition from the young-stand model to the established-stand model? And, how do we know when to effect this transition? Among the options are: 1) continue projecting the sample until the end of the planning horizon using the young-stand model; 2) only grow the sample using the established-stand model; 3) define some method by which we use the results from the young-stand model to initialize the established-stand model with some single cutoff age; or 4) examine another method by which we combine the results of the two inconguent models into a single set of results. Each method has advantages and shortcomings, which we examine in Section 8.3. In this section, we briefly introduce the two simulators.

8.1.1 Simulating Young Stands

The CONIFERS growth model, developed by Ritchie and Hamann (2006, 2008), provides an R package that is called rconifers (Hamann and Ritchie, 2009). The package includes two variants, or simulators, that can be used to project plant records. The default variant (SWO, variant #0) is used to project vegetation from southern Oregon and northern California. The alternative variant (SMC, variant #1) is used to project vegetation from the Pacific Northwest region of the United States. See Hamann and Ritchie (2009) for more detailed documentation. In this section, we present the basic functionality of the simulator, then introduce some functions used to process an rconifers sample.data object, then project sample.data objects forward in time, and finally generate summaries by species.

Data: field data, growth and yield equations, project objectives and constraint formulas, software
Result: Complete datasets of forecast future forest conditions
$S_{t=3} = $ sample.3;
$\hat{S}_{t=0} = ImputeMissingValues(S_{t=0})$;
$EvaluateSummaryStatistics(t = 0)$;
for $g \leftarrow 2$ **to** 100 **do**
$\quad S_{t+1} \leftarrow GrowOneYear(S_t)$;
$\quad Y_g = EvaluateSummaryStatistics(S_{t+1})$;
$\quad \mu_g \leftarrow UpdateParetoArchive(\lambda_g)$;
end
$ExportYields(Y_G)$;

Algorithm 1: Forest simulation process.

8.1.1.1 Loading the rconifers Package

We assume that the rconifers package is installed, for example using

```
> install.packages("rconifers")
```

We load the CONIFERS model using the `library` function and set the species map to the variant that we wish to use. Here we use variant 1.

```
> library(rconifers)
```

```
Initialized 19 functional species coefficients for variant # 0
coefficients version is 4.120000
```

```
> set.species.map(set.variant(1))
```

```
Initialized 3 functional species coefficients for variant # 1
coefficients version is 4.120000
```

The species map object, called `smc`, contains settings for three functional species-level coefficients that can be controlled by the user. We can examine it as follows.

```
> dim(smc)
```

```
[1]  3 13
```

```
> names(smc)
```

```
 [1] "idx"              "code"              "fsp"
 [4] "name"             "organon"           "cactos"
 [7] "fvs"              "endemic.mort"      "max.sdi"
[10] "browse.damage"    "mechanical.damage" "genetic.worth.h"
[13] "genetic.worth.d"
```

```
> head(smc)
```

```
  idx code fsp                 name organon cactos fvs
1   0   CV   0 Competing Vegetation       0      0
2   1   DF   1          Douglas fir     202      4  DF
3   2   NS   2          Non Stocked       0      0
  endemic.mort max.sdi browse.damage mechanical.damage
1        0.002       0             0                 0
2        0.002     450             0                 0
3        0.000       0             0                 0
  genetic.worth.h genetic.worth.d
1               0               0
2               0               0
3               0               0
```

or by using the `str` function.

Each functional species code is used to map an actual species into the co-efficient array for each variant. We use the term functional species to describe a group species of that behave similarly to one another in the growth model. The data frame object `smc` also contains a text species code that can be used in data recorder files. The species map also includes elements for mortality controls like endemic mortality (`smc$endemic.mort`), maximum stand density index values (`smc$max.sdi` Reineke, 1933), and values that can be used to control height growth like mechanical damage (`smc$mechanical.damage`) or damage caused by animal browsing (`smc$browse.damage`).

8.1.1.2 Creating and Using a sample.data Object

The rconifers package provides a specific class for storing the data to be used for simulation: a sample.data object. A sample.data object must contain at least two data frame objects: one to define plot-level attributes (called `sample.data$plots`) and one to define the tree records (called `sample.data$plants`). In additon to the data frame objects, the sample.data object must also include an age (called `sample.data$age`) and a parameter (called `sample.data$x0`) that defines the intercept term for a stand density trajectory (Reineke, 1933; Hann and Wang, 1990). Most of the rconifers functions operate directly upon sample.data objects.

First, load the plot and plant data frame objects from the package.

```
> data(plots.smc)
> data(plants.smc)
```

Next, create a list that contains the plots, plants, and initial age, and assign the resulting list a sample.data class,

```
> sample.3 <- list(plots = plots.smc,
+                  plants = plants.smc,
+                  age = 3,
+                  x0 = 0.0)
> class(sample.3)  <- "sample.data"
```

Now we can call many of the functions in rconifers directly upon the `sample.3` object. For example, to generate a set of species-level summaries, we use the summary function.

```
> summary(sample.3)

sample contains 18 plots records
sample contains 58 plant records
age =  3
x0 =  0
```

```
max sdi =  450
          qmd       tht        ba     expf
DF 0.578505 5.76724 0.456331 322.222
```

The `summary` function calls the `sp.sums` function, which generates a data frame object for which each row contains the summaries for a different species in the plant list. This data frame can also include non-tree species. The `summary` function also presents the `sample.data$age` and `sample.data$x0` variables.

8.1.1.3 Generating Young-Stand Simulations

We use the `project` function to project for a sample.data object. Here, we grow the `sample.3` sample.data object forward in time for 10 years, and simply print out the projected summary, using the `summary` function,

```
> summary(sample.3)

sample contains 18 plots records
sample contains 58 plant records
age =  3
x0 =  0
max sdi =  450
          qmd       tht        ba     expf
DF 0.578505 5.76724 0.456331 322.222

> sample.25 <- project(sample.3,
+                      22,
+                      control = list(rand.err = 0,
+                                     rand.seed = 0,
+                                     endemic.mort = 0,
+                                     sdi.mort = 0))
> summary(sample.25)

sample contains 18 plots records
sample contains 58 plant records
age =  25
x0 =  0
max sdi =  450
          qmd       tht        ba     expf
DF 8.59453 65.5306 129.815 322.222
```

Here the output includes the number of total plot records, the total number of plant records in the sample, the total age, the maximum stand density index (`max.sdi`, see Hann and Wang, 1990), the initial value at which `max.sdi` crossed the line of imminent mortality (Reineke, 1933), x0, and the output from the `sp.sums` function.

The `project` function includes a control argument that can be used to fine-tune the behavior of the function, and thus the simulator. `rand.err` is a flag to indicate that a normally distributed random number should be added to the height growth of the plant (`rand.err = 1`). `rand.seed` sets the initial value for the random number generator and only needs to be called once at the beginning of the simulation for each sample you want to project. If you need to repeat the randomness pattern, you need to reseed the random number generator with the same seed before projecting the sample. The value is not a switch; it requires an integer value to set the random number generator (see `?set.seed`). The `endemic.mort` argument is used to toggle the background mortality in the model applied to each growth cycle. Finally, the `sdi.mortality` switch is used to control additional mortality using the stand density index, defined by Reineke (1933). This last component of mortality is independent of the endemic mortality in the simulator. The influence of specific arguments is covered more completely in the package documentation (Ritchie and Hamann, 2006, 2008; Hamann and Ritchie, 2009).

As we described in Section 8.1, we ultimately want to compare simulations, given the same initial conditions. For example, in the following simulation run we assume that no vegetation control will be applied, and we generate simulations using a set of for loops to project a sample.data object for 100 years.

```
> res.v <- vector(length = 98, mode = "list")
> s0 <- sample.3
> res.v[[1]] <- data.frame(age = s0$age, sp.sums(s0)["DF",])
> ## grow from age 3 to age 100 (97 more years)
> for(m in 1:97) {
+    ## project s0 in one year intervals
+    s1 <- project(s0, 1, control = list(rand.err = 0,
+                                         rand.seed = 0,
+                                         endemic.mort = 1,
+                                         sdi.mort = 1))
+    res.v[[m+1]] <- data.frame(age = s1$age,
+                               sp.sums(s1)["DF",])
+    s0 <- s1
+ }
> res.v <- do.call(rbind, res.v)
> res <- res.v
```

In this example, the plant list in the sample s0 includes only Douglas-fir (*Pseudotsuga menzesii* Mirb. Franco), and so we can use the index operator to extract only the Douglas-fir results from the summaries generated by the `sp.sums` function (here `sp.sums(s1)["DF",]`). The process, as outlined in Section 8.1, is to assume that `sample.3` represents the initial conditions, compute the initial condition summary statistics, and then project s0 forward in time 1 year where the results are stored in s1. The process is repeated

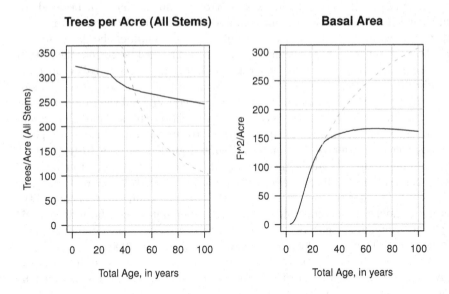

Fig. 8.1: Results from rconifers simulations using the SMC variant for the first 100 years of a simulation. These plots are for the `sample.data` object described in the text. The gray dashed line represents the trajectory from McArdle et al. (1949) (site index 180).

until the end of the planning horizon (here, 100 years). Figure 8.1 presents trajectories for two stand metrics: 1) trees per acre (from `sp.sums()$expf`) and 2) stand basal area (from `sp.sums()$ba`).

Had we included other functions from the rconifers package (for example, `thin`), the data frame `res` would also need to store those simulations so that the results would now be sufficient to examine the basic silvicultural relationships influenced by various initial conditions and silvicultural treatments (e.g., vegetation management or pre-commercial thinning). For a more complete presentation of all the functions in the rconifers package, the reader is refered to the rconifers documentation.

8.1.2 Simulating Established Stands

The CONIFERS model is a young-stand model and is suitable for projecting forest growth for about the first 25 years of growth. In order to provide projections after this age, we need an established-stand model. Our requirement is that the simulator can be run on multiple platforms, which means the source code must be available. We implement the empirical growth and yield tables for the Douglas-fir zone in Washington State developed by Chambers (1980). This approach allows us to provide a useful example of 1) building a model using external C or FORTRAN source code; 2) using that model from within R; and finally 3) integrating that model into our current project to create a uniform, and thus useful, stream of data.

The model developed by Chambers (1980) contains equations for predicting normal[1] basal area (nba), trees per acre for trees 7 inches and larger (tpa.7.plus), average stand diameter (qmd), and stand mean height for trees 5 inches and larger (smh). The model also includes equations for cubic foot volume, Scribner volume to a 6 inch top, Scribner volume in 16 foot logs, and Scribner volume in 32 foot logs using a tariff system. However, since rconifers does not provide volumes and we may want to compute volumes to some other specification, we defer volume determination until Section 8.2. In this section, we briefly present a subset of the source code, compile the source code for use within R, and finally load, attach, and call the functions to generate a data frame commensurate with the resulting data frame in Section 8.1.1.3.

8.1.2.1 Source Code

The files chambers-1980.h, chambers-1980.c, and chambers-1980.r contain equations and functions that define the growth and yield characteristics for the Douglas-fir (*Pseudotsuga menzesii* Mirb. Franco) zone in Washington State as defined in Chambers (1980). The header file (chambers-1980.h) defines the functions presented in the implementation file (chambers-1980.c), and we provide a set of R wrapper functions (chambers-1980.r).

In this section, we demonstrate the procedures for calling external functions using the equation to predict stand mean height (Chambers, 1980, Table 16), expressed as

$$\hat{H} = 2202.75000$$
$$+ 205.41618 \times (\log(A_{bh} \times S))^2$$
$$- 1323.88477 \times (\log(A_{bh} \times S))$$
$$- 0.00671 \times A_{bh}^2$$
$$- 383.35889 \times \frac{1}{A_{bh}}$$

where $\hat{H}$ is the predicted stand mean height for trees 5.0 inches and larger, A_{bh} is the age at breast height, S is the site index, defined by King (1966), and log is the base 10 logarithm.

---

[1] Normal is defined here informally as meaning typical or average for a large population of sampled stands.

The C function `chambers_1980_stand_mean_height` is defined in the file `chambers-1980.h` as

```
void chambers_1980_stand_mean_height(
    double    *site_index,
    double    *total_age,
    double    *stand_mean_height);
```

and the C implementation for the function is in the file `chambers-1980.c` as

```
void chambers_1980_stand_mean_height(
    double    *site_index,
    double    *total_age,
    double    *stand_mean_height)
{
    double   ret_val;
    double   bha;
    bha = chambers_1980_breast_height_age(site_index,
                                          total_age);
    if(bha < 1.0)
    {
        bha = 1.0;
    }
    ret_val =     2202.75000
        + 205.41618 * pow(log10(bha * (*site_index)), 2.0)
        - 1323.88477 * log10(bha * (*site_index))
        - 0.00671 * bha * bha
    - 383.35889 * (1.0 / bha);
    if(ret_val <= 0.0)
    {
        ret_val = 0.0;
    }
    *stand_mean_height = ret_val;
}
```

The function `chambers_1980_breast_height_age`, which is called from within the `chambers_1980_stand_mean_height` function, is also found in the file `chambers-1980.c`.

The function is called using three arguments: `*site_index`, `*total_age`, and `*stand_mean_height`. The first two arguments are the predictor variables. The last argument is a pointer to where the results are stored. All three arguments are pointers to variables of type `double` (that is, these arguments are passed by reference). At the time of this writing, all arguments to external functions or subroutines are passed by reference. A table of variable types is presented in Table 8.1.

Table 8.1: Foreign language data types.

R storage mode	C type	FORTRAN type
logical	int *	INTEGER
integer	int *	INTEGER
double	double *	DOUBLE PRECISION
complex	Rcomplex *	DOUBLE COMPLEX
character	char **	CHARACTER*255
raw	unsigned char *	none

8.1.2.2 Compile, Attach, Call, and Wrap External Code

The final step in the process of calling external source code is to 1) *compile* the source code into a shared library; 2) *attach* the resulting shared library to your current R process; 3) *call* the external code using one of the R functions for calling external libraries; and optionally 4) wrap the R external calling functions into a wrapper function of your own. The following code assumes that appropriate command-line tools are available to build R and R extensions.

Compile

Since the files chambers-1980.h and chambers-1980.c define and contain the functions themselves, we only need to issue a single command, from the operating system command prompt (here, $) to compile the shared library,

```
$ R CMD SHLIB chambers-1980.c
```

which will compile and link the code into a shared library chambers-1980.so (chambers-1980.dll in Windows). The call will present reasonably helpful error messages if necessary.

Attach

To attach the shared library and access the functions in the current R process, use the dyn.load function at the R prompt,

```
> dyn.load("chambers-1980.so")
```

upon which, unless something goes wrong, you should recieve no error message and be returned directly to the R command prompt.

Call

To call the function `chambers_1980_stand_mean_height`, which is written
in C, we must use either of R's `.C` or `.Call` function, passing the arguments
to the function as well,

```
> site.index <- 120.0
> total.age <- 60
> .C("chambers_1980_stand_mean_height",
+               as.double(site.index),
+               as.double(total.age),
+               smh = as.double(0))$smh
```

```
[1] 112.9387
```

which returns the correct value but looks cumbersome.

Wrap

For simplicity, we wrap the `.C` function call within a more accessible-looking
R function called `chambers.1980.smh`,

```
> chambers.1980.smh <- function(site.index,total.age) {
+   ret.val <- .C("chambers_1980_stand_mean_height",
+               as.double(site.index),
+               as.double(total.age),
+               smh = as.double(0))$smh
+   ret.val
+ }
```

which provides an easier interface to the original function and returns the
value of interest.

Finally, to verfiy the wrapper function is working properly, simply call the
R wrapper function `chambers.1980.smh`,

```
> chambers.1980.smh(120.0, 60.0)
```

```
[1] 112.9387
```

which matches the results presented in the original publication for a stand
with a 50 year site index of 120 feet at age 60 (Chambers, 1980, Table 16).

In all, there are about 14 functions defined in chambers-1980.h and
chambers-1980.c that need `.C` wrappers. We provide a set of wrapper functions so, to conserve space, the functions can be loaded by sourcing the file

```
> source("chambers-1980.r")
```

so that now all the functions required to produce the tables in Chambers (1980) from the functions defined in `chambers-1980.h`, `chambers-1980.c`, and `chambers-1980.r` are available from the R command prompt. For more details on calling external functions, we refer the reader to "Writing R Extensions", which is part of the R documentation.

8.1.2.3 Generating Established-Stand Simulations

We include another wrapper function, called `chambers.1980`, to project established stands,

```
> chambers.1980 <- function(ages = 1:100,
+                           site = 125.0,
+                           pnba = 1.0) {
+   ret.val <- matrix(0, length(ages), 5)
+   for(i in ages) {
+     res <- c(i,
+              chambers.1980.adbh(site, i, pnba), ## Table 5
+              chambers.1980.smh(site, i),        ## Table 16
+              chambers.1980.nba(site, i),        ## Table 1
+              chambers.1980.ntpa(site, i, pnba)  ## Table 3
+              )
+     ret.val[i,] <- res
+   }
+   ret.val <- as.data.frame(ret.val)
+   names(ret.val) <- c("age","qmd","tht","ba","expf")
+   ret.val
+ }
```

where the function returns a data frame object similar to that in Section 8.1.1.3 that includes our three critical metrics: 1) average stand diameter (`qmd`); 2) stand mean height (`tht`), and basal area (`ba`); and 3) trees per acre (`expf`) for each year in a projection. The results of a simulation are presented in Figure 8.2.

Figure 8.2, like Figure 8.1, contains plots of the number of trees per acre (`expf`) and the stand basal area (`ba`). The plots reveal a common phenomenon presented in forest simulation projects. The model by Chambers (1980) (gray line) contains only stems over 7 inches (17.78 cm), whereas the rconifers simulator (black line) reports values for all stems. The results of this disparity are clearly visible in both plots. These results suggest that the user needs to be aware of the input requirements, output limitations, and model assumptions for each simulator. We acknowledge this critical shortcoming and address it in Section 8.3. Regardless, we now have the ability to generate simulations by passing in a set of reporting ages, a site index (`site.index`), and a percent normal basal area (`pnba`) for *established stands*. We now need to consistently

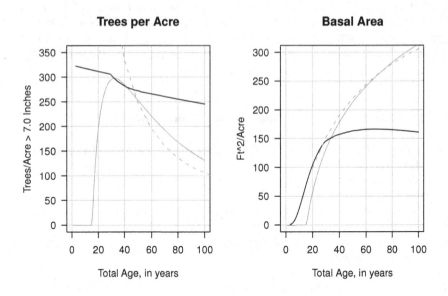

Fig. 8.2: Simulation results from the `chambers-1980` function where `site`=120 and `pnba`=1.0. The dark gray lines represent the resulting rconifers stand trajectories (see Figure 8.1).

generate volumes needed to determine the maximum mean annual increment and the volume distribution by log grade.

8.2 Generating Volumes

In this section, we generate log and stem volumes and calculate the distribution of stem volume by log grade for a set of simulations to examine our metrics of interest (specifically, total volume, mean annual increment, and volume distribution). To compute the volumes for our simulations, we use the taper function presented by Kozak et al. (1968) and obtain the merchantable height using the root-finding function `uniroot`. To simplify our task, we use the Smalian log rule (Briggs, 1994) and present the procedure in metric units. We then summarize the volumes by log grade for all merchantable stems in the young-stand simulations and, finally, compute the volumes for the established-stand simulations using the same method.

8.2.1 The Taper Function

The taper equation for Douglas-fir, developed by Kozak (1988), can be used to provide 1) predictions of inside diameter at any point on the stem; 2) estimates of total volume; 3) estimates of merchantable volume and merchantable height to any top diameter and from any stump height; and 4) estimates of individual log volumes. The taper equation function is

$$d_i = a_0 D^{a_1} a_2^{D} X^{b_1 Z^2 + b_2 \ln(Z + 0.001) + b_3 \sqrt{Z} + b_4 \exp Z + b_5 (\frac{D}{H})} \qquad (8.1)$$

where $X = (1 - \sqrt{h_i/H})(1 - \sqrt{p})$, D is the diameter at breast height (1.3 m), H is the total stem height, $p = 0.25$, $Z = \frac{h_i}{H}$, and h_i is the height above ground on the stem at which we want to determine d_i, the diameter inside bark, ln is the natural logarithm, and a_0, a_1, a_2, b_1, b_2, b_3, b_4, b_5 are estimated coefficients (Kozak, 1988).

Equation (8.1) can be written as a vectorized R function,

```
> bcmof.diDBH <- function(dbh, tht, cr, hi) {
+    p <- 0.25
+    dib <- 1.02453 * dbh^0.88809 * 1.00035^dbh
+    X <- (1.0 - sqrt(hi / tht)) / (1.0 - sqrt(p))
+    Z = hi / tht
+    a = 0.95086 * Z * Z;
+    b = -0.18090 * log(Z + 0.001);
+    c = 0.61407 * sqrt(Z) +  -0.35105 * exp(Z);
+    d = 0.05686 * (dbh / tht);
+    retval <- (dib * X^(a + b + c + d))
+    retval[tht < hi] <- 0
+    retval
+ }
```

where arguments for the function are *dbh*, *tht*, *cr*, and *hi*, which represent the diameter at breast height (1.3 m), the total stem height, the crown ratio, and the height at which we want to determine d_i, respectively. An example profile where *dbh* = 45 cm, *tht* = 27 m, and *cr* = 0.6 is displayed in Figure 8.3.

8.2.2 Computing Merchantable Height

To determine the height above the base of the stem at which we cannot make cuts (the merchantable limits), we need a function that will cross the *x*-axis at the merchantable height of the stem. A simple method to accomplish this is to subtract the merchantable diameter (md) from the diameter inside bark (diDBH)

```
> merch.height.func <- function(hi, dbh, tht, cr, md) {
+    dib <- bcmof.diDBH(dbh, tht, cr, hi)
+    diff <- dib - md
+    diff
+ }
```

so that when diDBH is less that md, the function crosses the x-axis; that is, the root of equation (8.1).

We pass the merch.height.func function to the uniroot function to obtain the merchantable height of the stem,

```
> mh <- uniroot(merch.height.func,
+                    c(0, 27),
+                    dbh = 45.0,
+                    tht = 27,
+                    cr = 0.60,
+                    md = 10.0)
> mh

$root
[1] 23.34488

$f.root
[1] 1.584768e-06

$iter
[1] 5

$estim.prec
[1] 6.103516e-05
```

where the arguments of the uniroot function include the limits of evaluation (c(0,27)) and any necessary arguments to pass through to the merch.height.func function (here dbh, tht, cr, and md).

The results of the uniroot function show that it took five iterations to arrive at the root (iter), which occurred at 23.34488 meters above the ground. To be precise, for the inputs we provided, the location of the root occured at 23.34488 and the taper function actually evaluates to 0.000002 (a zero, or root) at that value. The estimated precision after five iterations was 0.00006. Should we require greater precision, we can alter the tol and maxiter arguments to our satisfaction.

Finally, to verify the results, we can pass the resulting uniroot mh$root value back into the taper function to verify that we get the same value for the minimum merchantable diameter inside bark of 10.0 cm,

```
> min.dib.check <- bcmof.diDBH(45, 27, 0.60, mh$root)
> min.dib.check
```

Table 8.2: Log grades and merchandising specifications.

Sort name	Diameter range (inches)	Description
Pulp	$2 \leq D < 5$	Pulp logs (chipped)
Sawlog #3	$5 \leq D < 12$	Small sawlogs
Sawlog #2	$12 \leq D < 18$	Medium sawlogs
Sawlog #1	$18 \leq D < 32$	Large sawlogs
Peeler	$D \geq 32$	Veneer Logs

[1] 10.00000

Success. We may now proceed with summarizing the volumes by grade.

8.2.3 Summarizing Log Volumes by Grade

To generate the log volumes and volume distributions by grade for each time period, we must declare a log volume function and a set of vectors that contain the definitions for the minimum diameters for logs (log.breaks) and the names to assign to the grades (log.grades),

```
> smal.vol <- function(d1, d2, l) {
+    c <- 0.0001570796 ## in m^3
+    smal.vol <- (0.25 * d1^2 + 0.25 * d2^2) * l * c
+    smal.vol
+ }
> log.breaks=c(2,5,12,18,32,999)
> log.grades=c("pulp","s4","s3","s2","s1","peeler")
> grade.names <- c("Pulp","#4 Sawlog","#3 Sawlog",
+                  "#2 Sawlog","#1 Sawlog", "Peeler")
```

where the function smal.vol is the function for the Smalian volume (Briggs, 1994), $d1$, $d2$, and l are the small-end diameter, large-end diameter, and length for a log, in meters, and the log.breaks contain the smallest diameter for a log in the respective log.grades class in centimeters. The last entries in the log.breaks and log.grades vectors represent the largest category (see Table 8.2).

Equipped with these inputs and a tree list, we can proceed with merchandising our stems so that we obtain a vector of volumes, using log.grade, for each tree record. The process of merchandising each stem is as follows: 1) determine the merchantable height for the stem given the set of log.breaks; 2) compute the length of the stem that could be produced for each log grade; and 3) use a for loop to compute the volume for each log within the stem, with the length of each log being the difference in the heights between cuts in the stem at the minimum log diameters (at log.breaks).

First, we create a vector to store the merchantable heights for each of the
`log.breaks`. Then, we loop over the `log.breaks` vector to determine where
to make the cuts in the stem so that the stem is cut into lengths that represent
the longest possible log for each log grade,

```
> mh.bks <- rep(0, length(log.breaks))
> for(i in 1:length(log.breaks)) {
+    dbh <- 45.0; tht <- 27; cr <- 0.60
+    if(log.breaks[i] <= bcmof.diDBH(dbh, tht, cr, 0.0)) {
+      mh <- uniroot(merch.height.func,
+                    c(0, tht),
+                    dbh = dbh,
+                    tht = tht,
+                    cr = 0.60,
+                    md = log.breaks[i])
+      mh.bks[i] <- mh$root
+    } else {
+      mh.bks[i] <- 0.0
+    }
+ }
```

The resulting vector (`mh.bks`) contains the heights at which the stem can
no longer be merchandised into the associated `log.break` entry. For example,
the first entry in the merchantable height breaks (that is, `mh.bks[1]`), which
is

```
> log.breaks

[1]    2    5   12   18   32  999

> mh.bks

[1] 26.497448 25.405561 22.451219 19.407203   4.588263   0.000000
```

means that the longest log that can be cut from the stem while maintaining a small-end diameter larger than 2 cm is 26.49745 m above the ground.
Subsequently, the maximum length that can be cut from the stem for a 5 cm
log is 25.40556 m. Thus, the length of the top log is the difference between
the two lengths, 1.09189 m, so that the top log has a small-end diameter
(sed) of 2 cm, a large end diameter of 5 cm, and a length of 1.09189 m. The
Smalian volume given those dimensions is in cubic meters. Since the process
of iterating over tree records can be lengthy, we provide a function called
`generate.log.vols` so that given a sample.data object and a vector of minimum log diameters (`log.breaks`), the `generate.log.vols` function returns
a data frame that contains the original tree list, with volumes, by `log.grade`,
appended as columns. The function can be used to compute volumes for both
young-stand and older-stand model results.

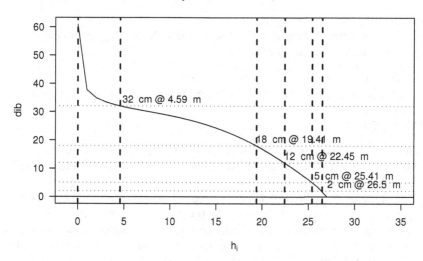

Fig. 8.3: The example stem with the bucking lines for logs defined by the `log.breaks` vector, given a stem with a breast-height diameter of 45 cm and a total stem height of 27 m.

Figure 8.3 presents the major features of merchandising a single stem using the methods described in this section. The vertical lines represent the crosscut decisions at each of the minimum height values (`mh.bks`). The horizontal lines represent the cylinders that would result from extending the small-end log diameter to the base of the stem (`log.breaks`). The profile of the stem diameter, at height h_i, is represented by the solid line.

8.2.4 Young-Stand Volumes

Using the `generate.log.vols` function, we can generate tree lists with volumes for each year in the young-stand simulations. Ultimately, we need to develop a single line, for each year, that contains the stand-level attributes (`qmd`, `tht`, and `expf`), the total volume (`sm.vol`), the volume by `log.grade`, and the distribution of that volume. Again, to reduce the code presented here, we have included another function in the file `generate_log_volumes.r`, called `sp.sums.2`. The function `sp.sums.2` performs all of the tasks described in this section and returns a set of summaries, by species, that includes the total log volume and the percentage of the volume in for the `log.grade` and log.breaks vectors. For example, to obtain the volumes for the young-stand simulation from Section 8.1.1.3 (that is, sample.25),

```
> ## the log.breaks are in cm
> sp2 <- sp.sums.2(sample.25, log.breaks/2.54, log.grades)
> sp2[,c("sm.vol", log.grades)]
```

```
     sm.vol      pulp       s4       s3       s2       s1
DF 3647.269  1.380405 21.88773 463.9883 1826.207 1321.220
     peeler
DF 12.58429
```

The resulting data frame object contains the quadratic mean diameter (qmd), the mean total height weighted by the stem's expansion factor (expf), the total basal area (ba), the total number of stems per acre (expf), the total number of stems per acre over 7 inches (expf.7.plus to compare our results against the Chambers (1980) model), the total Smalian log volume (sm.vol), a vector of Smalian volumes by log.grade, and a vector of proportions for the log volume (log.grades) for each species in sample.25. Here we only present the total volume and the volumes in each of the log.grades.

8.2.5 Established-Stand Volumes

To generate established-stand volumes using the function sp.sums.2, we need to create a temporary sample.data object from the results of the chambers.1980 function. The output from the established-stand simulator is a vector of attributes for any given age, so we can either 1) determine the volume from a single set of observations (that is, summary statistics like qmd and tht) or 2) generate a set of stem observations (a tree list) from the population of stems described by the summary statistics. We present both methods in this section.

8.2.5.1 From Summary Statistics

The first method is the simplest. Here we construct a single stem from the stand summary statistics from the established-stand simulator. For example, to determine the volumes for an established-stand simulation at age 50 years (e.g., site = 120, pnba = 1.0, age = 50), we first generate the established-stand simulation using the chambers.1980 function. Then, we create an rconifers plot and plant data frame objects and, finally, an rconifers sample.data object,

```
> ch <- chambers.1980(ages = 1:200, site = 120, pnba = 1.0)
> ch.stem <- data.frame(plot = 1,
```

```
+                          sp.code = "DF",
+                          d6 = NA,
+                          dbh = ch[50,]$qmd,
+                          tht = ch[50,]$tht,
+                          cr = 0.6,
+                          n.stems = 1,
+                          expf = ch[50,]$expf,
+                          crown.width = NA,
+                          errors = 0)
> ch.plot <- data.frame(plot = 1,
+                          elevation = NA,
+                          slope = NA,
+                          aspect = NA,
+                          whc = NA,
+                          map = NA,
+                          si30 = 120)
> ch.50 <- list(plots = ch.plot, plants = ch.stem, age = 50)
> class(ch.50) <- "sample.data"
> sh2 <- sp.sums.2(ch.50, log.breaks, log.grades)
> sh2[,c("qmd","ba","expf","sm.vol",log.grades)]
```

```
        qmd       ba     expf   sm.vol     pulp        s4       s3
DF 12.13548 207.8756 258.7996 9032.07 22.49502 360.6965 8481.45
            s2 s1 peeler
DF 167.4276  0      0
```

Note that the volumes are generated from a single representative stem for the stand.

8.2.5.2 From Tree Lists

There are many techniques for generating tree lists. Here we generate a normal distribution of diameters. Using the normal function, and choosing year 50, we generate round(ch[50,]$tpa) realizations from an assumed normal population of stem sizes,

```
> dia.obs <- rnorm(n = round(ch[50,]$expf)*10,
+                  mean = ch[50,]$qmd,
+                  sd = 0.20*ch[50,]$qmd)
```

where mean is the quadratic mean of the tree diameters, sd is the standard deviation of the population which we assume is 20% of *qmd*, and ch[50,] is the data for the 50th year in the simulation. To increase the number of stems in our sample, say to generate a smoother distribution of stems, we would generate ten times the number of stem observations and divide the expansion

factor for each entry by ten to obtain the correct total number of stems per
unit area.

To construct the tree list, we include a set of values for the heights. Here
we use the equation to predict total height given DBH, presented by Hanus
et al. (1999), to generate the heights for each of the diameter realizations,
and assume a crown ratio of 40% (that is, cr = 0.40) and a single plot
representing the sample,

```
> ## generate the plant records
> plant.obs <- data.frame(plot = 1,
+                         sp.code = "DF",
+                         d6 = NA,
+                         dbh = dia.obs,
+                         tht = 4.5 + exp(7.262195456 +
+                                  -5.899759104 *
+                                  dia.obs^-0.287207389),
+                         cr = 0.40,
+                         n.stems = 1,
+                         expf = 1/10,
+                         crown.width = NA)
> plot.obs <- data.frame(plot = 1,
+                        elevation = 1000,
+                        slope = 0,
+                        aspect = 0,
+                        whc = NA,
+                        map = NA,
+                        si30 = 85.0)
> ch.50 <- list(plots = plot.obs,
+               plants = plant.obs,
+               age = ch[50,]$age,
+               x0 = NA)
> class(ch.50)  <- "sample.data"
```

Finally, we can use the modified function sp.sums.2 to obtain the metrics
of interest:

```
> sh2 <- sp.sums.2(ch.50, log.breaks, log.grades)
> sh2[, c("qmd", "ba", "expf", "sm.vol", "log.grades)]

        qmd       ba expf   sm.vol    pulp       s4       s3
DF 12.43444 218.4129  259 7868.594 20.05454 328.2135 6692.396
        s2       s1 peeler
DF 819.5235 8.406364      0
```

methods, as in Section 8.2.4.

Finally, when we no longer need the chambers-1980.so shared library, we
unload the library using the dyn.unload function,

```
> dyn.unload("chambers-1980.so")
```

8.3 Merging Yield Streams

For each of the two models, we can now produce data frame objects that contain projected summaries that include volumes and volume distributions for each of the log grades. Before we can examine our results, however, we must address the dilemma presented in Section 8.1.2.3. We need to provide a smooth transition between young-stand and established-stand simulations. Here our solution is to construct a linear combination similar to that presented by Wykoff et al. (1982), which creates results so that values from the young-stand simulations are used at young ages followed by some mixture of the two, and then, as the simulation progresses, the established-stand model defines the characteristics of the stand.

A linear combination of the values from the two data frames for each simulator can be expressed as

$$S_{\alpha,t} = \alpha(t)S_{y,t} + (1 - \alpha(t))S_{e,t} \qquad (8.2)$$

where $S_{\alpha,t}$ is the linear combination of $S_{y,t}$, the young-stand simulation variable at time t, $S_{e,t}$, the established-stand simulation variable at time t, and $\alpha(t)$ is defined by

$$\alpha(t) = \begin{cases} 1 & \text{for } 1 < t < 20 \\ \frac{t-t_s}{t_e-t_s} & \text{for } 20 \le t \le 40 \\ 0 & \text{for } t \ge 40 \end{cases} \qquad (8.3)$$

where $t_s = 20$ is the starting age at which the transition between the two models begins and $t_e = 40$ is the end of the transition period. Here, when $t < 20$, the young-stand model defines the yields. When $t > 40$, the established-stand model defines the yields. For example, at age 35, the value of the statistic is a linear combination of 25% young-stand model and 75% established-stand model. For other situtations, the starting and ending values of the transition period may be substantially different. This method should not be applied to ordinal or categorical data.

8.4 Examining Results

Figure 8.4 presents a comparison of the results for the individual models. The results for the stems per acre (expf) and basal area (ba) appear commensurate with the results from McArdle et al. (1949). The trajectories for mean stand height (tht) and, as a result, volume (sm.vol) appear patho-

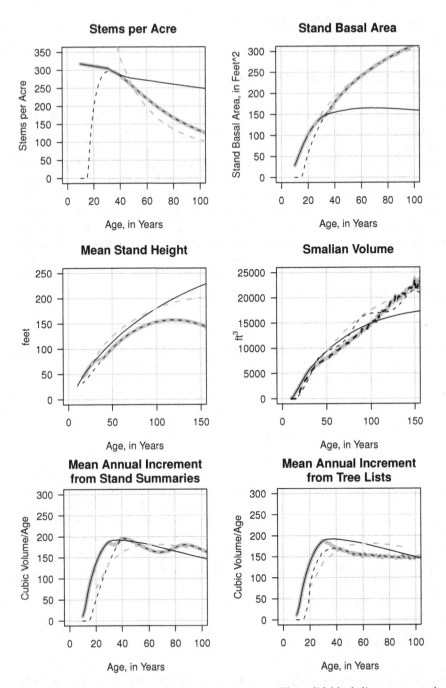

Fig. 8.4: Common metrics for forest simulations. The solid black lines present the results from rconifers. The dashed black lines are the results from the chambers-1980.so simulator. The thick gray solid line presents the linear combination, defined using equation (8.2). The dashed gray line presents values from McArdle et al. (1949) for a normal stand with the site index of 180.0 feet. The thick dashed line in the Smalian volume represents the volume generated using the method described in Section 8.2.5.2.

logical, however, when compared with both McArdle et al. (1949) and the young-stand simulator rconifers. The resulting stand height, as described in Chambers (1980), exhibits negative height growth; we can interpret this as a sign that the established-stand simulator does not extrapolate well beyond the intended domain of the original data ($20 < $ age $ < 100$).

The Smalian volume plot presented next to the mean stand height in Figure 8.4 exhibits an artifact of the height function behavoir as well. The thin dashed gray line represents the volume computed from a single tree record (see Section 8.2.5.1). The line exhibits a sinusoidal nature with increasing amplitude. This artifact is corrected by generating a tree list using the stand-level attributes from the established-stand model (see Section 8.2.5.2). The final results (thin black dashed line in the thick gray solid line) represent the volumes computed from the generated tree lists using the method presented in Section 8.2.5.2. Note that this volume trajectory maintains a realistic path: the line exhibits a roughness resulting from the generation of stems independently at each year in the simulation. Finally, the mean annual increment trajectories, at the lower-left of Figure 8.4, include the resulting Smalian volumes from the methods presented in Sections 8.2.5.1 (uncorrected) and 8.2.5.2 (corrected), respectively, and also exhibit behavior similar to that found in the volume plot.

To determine the maximum mean annual increment, we obtain the maximum of the mean annual increment, using the which.max function to identify which row to retrieve from the data frame for each model,

```
> a.max.mai <- a[which.max(a$mai),]
> b.max.mai <- b[which.max(b$mai),]
> b.prime.max.mai <- b.prime[b.prime$mai == max(b.prime$mai),]
> c.max.mai <- c[c$mai == max(c$mai),]
> c.max.mai$col <- "c"
> c.prime.max.mai <- c.prime[c.prime$mai == max(c.prime$mai),]
> c.prime.max.mai$col <- "c.prime"
> max.mais <- rbind(a.max.mai,
+                   b.max.mai,
+                   b.prime.max.mai,
+                   c.max.mai,
+                   c.prime.max.mai)
```

where a represents the young-stand rconifers model, b represents the results from the established-stand simulator, b.prime contains the corrected volume results (see Section 8.2.5.2), c contains the uncorrected volume results, and c.prime contains the combined corrected results.

We now have all the data required to examine the influence of our different fusion methods on the maximum mean annual increment. For example, Table 8.3 presents the major metrics of interest, including the total volume, age, and tree density (TPA). For this table, we can examine basic summary statistics such as the ranges in age, variance in the basal areas, and aver-

Table 8.3: Statistics for simulations at the maximum mean annual increment for each of the models.

Simulation	Age	QMD (in)	Height (ft)	Basal Area (ft^2)	TPA	TPA 7"+	Volume (ft^3)
1	38	10	93	153	290	285	7305
2	40	11	82	178	285	285	7863
3	35	10	71	165	295	274	5967
4	40	11	82	178	285	285	7863
5	30	9	77	142	305	300	5626

age volume at maximum MAI. Here we only present our summary results to demonstrate the method.

In the next two sections, we further examine our results, focusing specifically on the volume and the mean annual increment, and finally answer our original questions: 1) What, if anything, happens to the distribution of log product volumes over time; and 2) What, if anything, happens to the resulting maximum mean annual increment values?

8.4.1 Volume Distribution

The top row of Figure 8.5 contains the smallest two log grade classes in the simulations (pulp and #4 sawlog). The results from the young-stand model increase rapidly below age 20 for both. After age 20, the pulp grade continues to increase, but more slowly, and the #4 sawlog volume also continues to increase after a slight dip. The uncorrected combined results (solid gray line) from the established-stand simulator increase rapidly, peak, and then also decline, but in a smooth line rather than showing the extreme changes exhibited by the young-stand simulator.

Medium sawlog grades (#2 and #3 sawlog) are displayed in the middle row of Figure 8.4. The results for the young-stand simulator increase monotonically over the entire range of the simulation, whereas the uncorrected combined results (light-gray dashed line) peak and decline near the middle of the simulation. The corrected combined results (solid and dashed black lines in thick gray lines) also peak, albeit lower and slightly sooner than the uncorrected combined results. Likewise, the results for the #2 sawlog grade appear to exhibit the same results, but later in the simulation. Also noteworthy is the fact that the volume for the uncorrected combined results begins to increase later, peaks higher, and crosses the corrected combined results. Our simulations do not extend far enough in time to determine when the two trajectories cross again.

At the bottom of Figure 8.5, the results for the combined corrected model begin to exhibit Peeler grade volume shortly after age 100 years. The de-

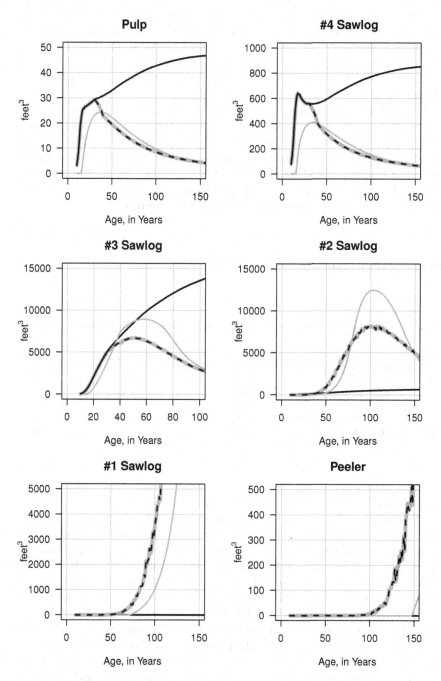

Fig. 8.5: Volume development, for each log grade, over the planning horizon. The young-stand simulator results are represented by a thin black solid line. The uncorrected established-stand simulation is represented by a thin gray solid line. The thick gray line with the thick dashed black line inside represents the merged young-stand and corrected established stand simulations.

velopment of #1 Sawlog volume follows a similar pattern, with the stand age around 60 years for the corrected combined results (black dashed line inside thick gray solid line), followed 20 years later by accumlated volume in the uncorrected combined results. When we examine ages above our original range (150 years), only then do any of the simulators produce volumes in the largest class.

The patterns and resulting conclusions we can infer from Figure 8.5, similar to those in Figure 8.4, are that 1) the method used to merge models can greatly influence the behavoir of the resulting combination; 2) the resulting simulations, and resulting combinations, may exhibit behavoir similar to published results (Chambers, 1980); and 3) producing results using a small sample (Chambers uncorrected) can lead to peculiar behavoir, and the decisions resulting from those simulations should be viewed critically.

8.4.2 Mean Annual Increment

Table 8.4: Volume metrics for simulations at maximum mean annual increment.

Model	Age	Pulp	#4 Sawlog	#3 Sawlog	#2 Sawlog	#1 Sawlog	Peeler	Total	MAI
1	38	31	560	6599	115	0.000	0	7305	192
2	40	24	397	7438	3	0.000	0	7863	197
3	35	24	431	5421	91	0.003	0	5967	170
4	40	24	397	7438	3	0.000	0	7863	197
5	30	29	559	4986	52	0.000	0	5626	188

Mean annual increment, which determines the biological rotation age in many forests (Leuschner, 1990; Davis et al., 2001), can be greatly influenced by the results of the growth and yield models used to compute the volumes over time. Table 8.4 presents the age, the Smalian volumes (see Section 8.2) by log.grade, and the mean annual increment at the age when maximum mean increment occurs. This is the time point that corresponds with the highest average yield. It is important to note that Table 8.4 represents only a single moment in time: when the mean annual increment is maximum.

The range in values in projected MAI, from a low of 5966.92 ft^3 to a high of 7863.18 ft^3, is reasonably compact. Conversely, the range in ages when the maximum occurs is considerable when compared with the values themselves. The ages range from a low of 35.00 ft^3 to a high of 40.00 ft^3. This range may have a considerable effect on the projections of forest- or landscape-level services.

8.5 Exporting Yields

To export the results from the simulations (`res$rconifers`) into a database program, we could use an SQL file. First, we would construct a loop to generate SQL INSERT commands and then write the results to a file named `smc-yields.sql`:

```
> hres <- res$rconifers
> names(hres)[3] <- "rx"
> sql.file <- file("smc-yields.sql", "w")
> sql.columns <- names(hres)
> for(i in 1:nrow(hres)) {
+    sql.command <-
+        sprintf("insert into results (%s) values (%s);",
+                paste(as.vector(sql.columns), collapse=","),
+                paste(hres[i,], collapse=","))
+    cat(sql.command, file = sql.file, sep = "\n")
+
+ }
> close(sql.file)
```

8.6 Summary

Forest simulation is a complex subject. Depending on the level of resolution required in the analysis (tree, stand, or forest), the process may be as simple as a single equation (model component) or require the development of a shared library or simulation package (chambers-1980.so and rconifers) and include multiple software packages (R and SQL), which could include using the **system** function to run command-line applications (FVS, ORGANON, CACTOS).

In this chapter, we used R to generate simulated yield tables, projections, or *sets of possible future forest conditions* that could be analyzed within R or exported to other packages (for example, a harvest scheduler). Our methods, while basic, provide a starting place for more complex analysis techniques.

Chapter 9
Forest Estate Planning and Optimization

9.1 Introduction

So far, we have developed procedures for processing field data, producing complete inventories, and generating forecasts of future forest conditions. In this chapter, we use R to solve a forest estate planning problem assuming we already have a sufficient inventory to describe the forest. Here we present and solve a linear programming (LP) problem that was originally presented by Leuschner (1990, see Chapter 4). The objective is to determine the harvest schedule for a classical method of forest regulation, the strict area harvest schedule. Strict area control is an indirect method of controlling the amount of forest produced in each cutting period. To achieve our objective, we use the glpk package (Lee and Luangkesorn, 2010), which is the R interface to Andrew Makhorin's GNU linear programing kit (GLPK) (Makhorin, 2009).

This problem provides a gentle introduction to mathematical programming that does not require extensive background development. There are other types of harvest schedules, model formulations, and constraints that can be applied to develop more interesting problems, but they are beyond the scope of this book. For more complete presentations of forest estate planning problems, we recommend Leuschner (1990), Rönnqvist (2003), and Bettinger et al. (2009).

The chapter is structured as follows. In Section 9.2, we briefly present a basic formulation framework. In Section 9.3, we present the tasks that are required to translate the model formulation into the components that are required to solve the formulation, namely 1) generating the decision variable columns, 2) the objective function coefficients, 3) row constraints, and 4) forest-level row constraints. In Section 9.3.7, we then obtain a solution given the developed model. Then, in Section 9.3.8, we extract the solution, decode the decision variables, compute forest-level metrics, and examine important features of the solution. Finally, we generate a machine-readable archive of the problem in Section 9.3.9.

9.2 Problem Formulation

Our first objective is to formulate the forest estate planning problem. While the formulation for each problem is unique, here we present a general formulation and notation that can be applied to many different problems. Here we use notation commonly found in computer science, operations research and management sciences. For each problem, we assume that the forest estate plan is to cover T planning periods. The length of each of the planning periods is ℓ years, and the total length of the planning horizon is denoted as $|T|$ years, where $T \times \ell = |T|$.

Let polygon $\mathscr{A} \subset \mathbb{R}^2$ represent a forest with a finite boundary, where the surface area, measured in acres[1], is denoted $\Lambda(\mathscr{A})$. The polygon $\mathscr{A}$ is partitioned into A polygons (stands). Each stand polygon, denoted $\mathscr{A}_b$, is defined by a unique label $\mathscr{B} = \{1, \ldots, b, \ldots, A\}$. The total area of each stand, measured in acres, is denoted $\lambda(\mathscr{A}_b)$, where $\bigcup_{\mathscr{A}_b} = \mathscr{A} \ (\mathscr{A}_b \subset \mathbb{R}^2)$, and $\bigcap_{\mathscr{A}_b} = \emptyset \ \forall \, b \in \mathscr{B}$. Let $\lambda(\mathscr{A}_b)$ denote a vector of length A representing the areas of $\mathscr{A}_b$ such that $\sum \lambda(\mathscr{A}_b) = \Lambda(\mathscr{A})$.

Let $\mathbf{Y}$ represent an $A \times T$ matrix of initial and future conditions (per unit area yields) for each stand $\mathscr{A}_b$ for each period $\ell \in T$ (*the planning horizon*).

Let $\mathscr{A}_b^{\mathbf{x}}$ represent the decision vectors for each stand $\mathscr{A}_b$, where $\mathbf{x}$ is a vector of length T. The T elements of $A_b^{\mathbf{x}}$ represent the area harvested from $\mathscr{A}_b$ over the planning horizon, subject to $\mathscr{A}_b^{x_t} \leq \lambda(\mathscr{A}_b)$ for each period t. This constraint ensures that *at most* the total area of each stand is harvested in each period. To conserve notation, $\mathscr{A}_b^{\mathbf{x}}$ will be represented by $\mathbf{x}$, where $\mathbf{x}$ is an $A \times T$ length vector of decision variables

$$\mathbf{x} = (\mathscr{A}_1^{\mathbf{x}}, \ldots, \mathscr{A}_b^{\mathbf{x}}, \ldots, \mathscr{A}_A^{\mathbf{x}}) \tag{9.1}$$

where $\mathbf{x} \in \mathbb{B}, \mathbb{I}, \mathbb{R}$, or some combination thereof.

Finally, given the forest $\mathscr{A}$ and a set of initial and future conditions $\mathbf{Y}$, a forest estate optimization problem can be expressed using the general problem formulation

$$\mathbf{x}^* = \underset{\mathbf{x} \in \mathscr{S}}{\operatorname{argmax}} \left\{ f(\mathbf{x}) \in \mathbb{R} \, \middle| \, \mathbf{g}(\mathbf{x}) \geq \mathbf{0}, \mathbf{h}(\mathbf{x}) = \mathbf{0} \right\} \tag{9.2}$$

where the goal is to obtain the arguments $\mathbf{x}^*$ that yield the maximum f, which is the objective, subject to $\mathbf{g}$, which is a vector of K inequality constraints; $\mathbf{h}$, which is a vector of M equality constraints; and $\mathscr{S}$, which is some subset of $(\mathbb{I}, \mathbb{N}, \mathbb{R})$ or some combination thereof.

When f, $\mathbf{g}$, and $\mathbf{h}$ are linear in the parameters, then equation (9.2) can be expressed as a *linear programming* problem.

We refer the reader to Dantzig (1963) for more details on linear programming, Bettinger et al. (2009) for more examples and formulations, and Makhorin (2009) for specifics on the glpk package.

---

[1] We retain the units in which the problem was originally expressed.

9.3 Strict Area Harvest Schedule

For this problem, our objective is to generate a solution for a strict area harvest schedule that maximizes the total woodflow of $\mathscr{A}$ over a planning horizon of six 10 year periods ($T = 6, \ell = 10, |T| = 60$ years) subject to some set of physical and policy constraints. Specifically, suppose that we have a forest with a finite boundary, denoted $\mathscr{A}$, and the function for the surface area, denoted $\lambda(\mathscr{A})$, returns 84000 acres. Also, $\mathscr{A}$ is partitioned into eight stand polygons (so $A = 8$), where the areas of the stands are

$$\lambda(\mathscr{A}_b) = (5000, 5000, 5000, 5000, 30000, 20000, 12000, 2000)$$

Since these values are relatively easy to keep track of using atomic objects, we begin by simply assigning them to individual variable names,

```
> A <- 8      # A
> T <- 6      # T
> l <- 10     # \ell
> P <- T * l # |T|

> ## \lambda(\mathcal{A}_{b})
> stand.acres <- c(5000,5000,5000,5000,30000,20000,12000,2000)
```

For this example, we will use the published yields from the original problem as presented by Leuschner (1990, see Chapter 4, Table 4.2). Here we have entered the values into a text file (located at ../../data/leuschner.txt), which can be read in using the **read.table** function,

```
> leusch.ylds <- read.table("../../data/leuschner.txt",
+                           header = TRUE)
```

where the resulting **leusch.ylds** object contains four columns: 1) the stand identifier (**stand** is the label for $\mathscr{A}_b$); 2) the planning period (**per**); 3) the age of the stand ($A_b^a = $ **age**); and 4) the yield (*volume*) ($\mathscr{A}_b^v = $ **vol**) in thousands of cubic feet per acre if we harvest stand $\mathscr{A}_b$ in period t.

To solve this problem, we need to generate the 14×48 detached coefficient matrix that stores the coefficients for the problem. The results matrix should have a set of eight rows that define the area constraints, a set of six rows that define the policy constraints, and 48 columns for the decision variables.

To generate all the required components to solve this problem, we first need to load the glpk package and create a glpk problem object,

```
> library(glpk)
> leusch.lp <- lpx_create_prob()
```

Next, we assign a symbolic name to the problem object:

```
> lpx_set_prob_name(leusch.lp, "leuschner")
```

By default, glpk will *minimize* the objective function. Since we want to *maximize* the objective function (equivalently, $-\min(-f)$), we need to call the `lpx_set_obj_dir` function to change the direction of optimization,

```
> lpx_set_obj_dir(leusch.lp, LPX_MAX)
```

where `GLP_MAX` indicates maximization.

The glpk problem object we have now created (called `leusch.lp`) is the object that stores the information about the optimization problem in memory and allows us to pass information to the glpk library directly, and almost all functions in the glpk package require it.

9.3.1 Objective Function

For this problem, the objective is to maximize the total yield

$$f = \mathscr{A}_b^{v_t} \ \forall \, t \in T, b \in \mathscr{B} \tag{9.3}$$

where the objective function (*total yield*) f is the sum of the product yields from stand $\mathscr{A}_b^v$, harvested in period t, over the entire forest $\mathscr{A}$. In this problem, the vector of yields for each stand ($\mathscr{A}_b^v$) is known exactly (that is, *non-stochastic*), so the only decision variables required ($\mathscr{A}_b^x$) represent the number of acres to harvest in $\mathscr{A}_b$ in period t (Marti, 2005).

In the following sections, we add columns, give them names, and place bounds on the columns.

9.3.2 Adding Columns

To add columns (*decision variables*) to the newly created problem object, we now use the `lpx_add_cols` function. For this problem, the length of **x** is 48, and so we need to add 48 columns to hold the 8 stand polygons × 6 harvest periods:

```
> lpx_add_cols(leusch.lp, nrow( leusch.ylds ))
```

```
[1] 1
```

where the 1 is the response from the glpk library. In this case, it is the ordinal number of the first new column added to the problem object.

9.3.3 Naming Columns

To generate names for the columns, we use the **paste** function with the
stand and **period** columns of the **leusch.ylds** data frame object to generate
human-readable labels for each decision variable,

```
> leusch.ylds$dv <- paste("a",
+                         leusch.ylds$stand,
+                         leusch.ylds$period,
+                         sep="")
> head(leusch.ylds)

  stand period age  vol  dv
1    1      1    1   30  3.2 a11
2    1      2   40  6.1 a12
3    1      3   50  8.3 a13
4    1      4   60 10.1 a14
5    1      5   70 11.6 a15
6    1      6   80 12.9 a16
```

where **leusch.ylds\$stand** $= b$ and **leusch.ylds\$per** $= p$ (see equation (9.3)).

The resulting column labels (**leusch.ylds\$dv**) are easy to read for this
problem. For example, the label **a13** can be deciphered as "harvest stand 1 in
period 3". The reason for the **a** character is that many solvers require the first
character of the input label to be non-numeric. There may be constraints on
the length of the label that can be used for each different solver, and so, for
larger problems (e.g., $A > 100000$ or $T > 100$), this simplistic approach might
not work. For example, if this problem contained over $10,000,000$ stands,
there would be no room left in a label that could only contain eight characters using the method (that is, **nchar("a10000000")** = 9 is too long). For
specific formats and naming conventions, we recommend that you consult
the documentation for the solver of interest. Here, however, we can use this
method because our solution space is sufficiently small.

Finally, to assign names to the columns, use the **lpx_set_col_name** function and loop over all of the entries in the **leusch.ylds** data frame object,
passing the newly created label to be used as the column name to the function,

```
> for( t in 1:nrow(leusch.ylds)) {
+    label <- leusch.ylds[t,]$dv
+    lpx_set_col_name(leusch.lp, t, label)
+ }
```

9.3.4 Bounding Columns

To set the bounds on the columns (*decision variables*) so that the harvested area from each stand, in any period, is zero or greater ($A_b^x \geq 0 \ \forall t \in T$),

```
> for(t in 1:nrow(leusch.ylds)) {
+   lpx_set_col_bnds(leusch.lp, t, LPX_LO, 0.0, 0.0)
+ }
```

The `lpx_set_col_bnds` function takes as its arguments the lp object, the row number, the type of bound being set (LPX_LO), and a set of bounds. Here we only need one bound since we are setting a lower bound only. For more details on setting bounds, see the glpk documentation (specifically, `?lpx_set_col_bnds`).

9.3.5 Setting Objective Coefficients

Finally, to set the objective function coefficients described in Section 9.3.1, we use the `lpx_set_obj_coef` function,

```
> for(t in 1:nrow( leusch.ylds)) {
+   ## set the objective coefficient in period t = 0.0
+   lpx_set_obj_coef(leusch.lp, t, leusch.ylds[t,]$vol)
+ } #$
```

We now have the major components to define our decision variables, the bounds on those variables, and the objective function coefficients. Next, we add row contraints for each of the stand areas (these are *physical constraints*) and a constraint for the forest as a whole (a *policy constraint*).

9.3.6 Adding Constraints

Constraints are commonly classified into several types (Weintraub and Navon, 1976; Davis et al., 2001; Bettinger et al., 2009). To simplify our presentation, we classify our constraints into physical constraints (e.g.,land availability, accessibility to transportation, or no ground equipment on ground over 60% slope) and policy constraints (e.g.,risk avoidance or maximum opening size, or only one assigned schedule per stand).

This problem has both types of constraints:

1. One or more physical constraints limiting the total forest area available for harvest.

2. A policy constraint to control cutting so that the acres harvested over all periods must equal the total acres of the forest (i.e., strict area control).

Since the goal is to turn the current forest $\mathscr{A}$ into a well-regulated forest, where an equal area will be harvested in each period, we need to include two physical constraints,

1. stand total area constraints and
2. total harvest area constraints

so that we can generate the final volume flow policy constraints. We first add the rows for each stand area and then add the constraints for the strict area control.

9.3.6.1 Stand Area Constraints

An unconstrained problem might harvest an infinite number of acres in each period, but since the total number of acres available for harvest in any period is limited to the total area of the forest ($\Lambda(A) = 84000$), we must constrain the total harvested area to be no larger than the total forest area (84000 acres). Mathematically, this constraint is expressed as

$$\sum_t \mathscr{A}_b^{x_t} \leq \lambda(\mathscr{A}_b) \; \forall \, b \in \mathscr{B} \tag{9.4}$$

where $\mathscr{A}_b^{x_t}$ is the total available acres in stand b harvested in period t, and $\lambda(\mathscr{A}_b)$ is the total available acres in stand b.

Since each stand may be harvested in more than one period, we must ensure that the total area that is harvested is no larger than the total stand area over all periods. To accomplish that, we must add a set of rows for each period in the planning horizon and include a sum of all the acres harvested for each stand over all periods.

First, add the constraint rows, using the `lpx_add_rows` function to create a row for each stand,

```
> lpx_add_rows(leusch.lp, length(stand.acres))
```

`[1] 1`

Create a matrix to contain sets of coefficients for each row, using $I \otimes 1$, where I is an $A \times A$ matrix, 1 is a column vector of length P (*number of periods*), and $\otimes$ is the Kronecker product,

```
> acre.consts <- t( kronecker(diag(length(stand.acres)),
+                     as.matrix( rep(1,6))))
```

The result is an 8×48 matrix called `acre.consts`. Now set the row names, bounds, and row matrix coefficients for $\mathscr{A}_b$ using `lpx_set_row_name`, `lpx_set_row_bounds`, and `lpx_set_mat_row`, respectively.

To do that, we create a for loop as we did for the columns in Section 9.3.6, generate the entries for the row, and finally set the row name, bounds, and coefficients. Here we perform all three functions at the same time rather than creating a loop for each operation as we did previously.

First, we create a variable to store the coefficient values for this portion of the detached coefficients matrix. The variable, val, is a vector of ones that holds the coefficients

```
> val <- rep(1, 6)  # this is the value of the constraint
```

Then, we loop over the stands to create the row coefficients for the number of acres harvested over the six periods. Since we can create all six entries for each of the six columns that need to be set for each row, we create a vector that will contain the column index (idx) by creating a loop within the outer loop, to generate the values for idx, and then perform the row operations,

```
> ## loop over the stands to generate
> ## the idx is the index for the set of rows
> ## we are trying to set the coefficients for
> for(i in 1:length(stand.acres)) {
+
+   idx <- rep(0, 6)  # this is the index of the col coeffs
+
+   ## manually create the index using id
+   ## and assign row i, column j stored as idx[id]
+   id <- 1
+   for(j in 1:ncol(acre.consts)) {
+     if(acre.consts[i,j] != 0.0) {
+       idx[id] <- j
+       id <- id + 1
+     }
+   }
+
+   ## now, perform all three tasks within the same loop
+   ## set the constraint row name
+   lpx_set_row_name(leusch.lp, i, paste("s", i, sep=""))
+
+   ## set the upper bound on the acres for this stand
+   ## to make sure no more than stand.acres[i] are cut
+   lpx_set_row_bnds(leusch.lp, i, LPX_UP, 0.0,
+                       stand.acres[i])
+
+   ## set matrix row coefficients (?glp_set_mat_row)
+   ## idx = ???, row index = i, val = vector of 1's
+   lpx_set_mat_row(leusch.lp, i, length(val), idx, val)
+ }
```

By now, we should have added eight and eight columns to the problem object. To check, we can use the `lpx_get_num_rows` and `lpx_get_num_cols` functions to obtain the number of rows and columns in the problem object,

```
> lpx_get_num_rows(leusch.lp)
```

`[1] 8`

```
> lpx_get_num_cols(leusch.lp)
```

`[1] 48`

The output verifies our results.

Note that in Sections 9.3.1 through 9.3.5 we used separate loops to present the glpk functions. Here we combined loops to conserve space, but it is important to verify that the matrix generation procedures you use create the matrix you intend. Without verification of your code, you might inadvertently solve another problem, as mathematical programming solves the problem you actually specify, not the problem you intended to specify.

9.3.6.2 Strict Area Control

In a classic regulated forest, the acres are distributed so that there are $\frac{\Lambda(\mathscr{A})}{T}$ acres in each age class from 1 to T periods. A strict area control schedule requires that the forest be regulated so that the maximum woodflow occurs when the area weighted mean annual increment (MAI) is maximum. Leuschner (1990) determined that the maximum MAI, for $\mathscr{A}$, occurs at 60 years, given the area weighted mean base age 50 year site index is 113.57 feet. Given that data, the annual harvest in hundreds of cubic feet should be

$$\sum_j a_{ij} = \frac{\Lambda(\mathscr{A})}{r} = \frac{84000}{60} = 1400 \qquad (9.5)$$

where r is the rotation age in years (e.g., 60) (Leuschner, 1990, p. 33). The result should be multiplied by 10 to obtain the harvest per planning period ($\ell = 10$).

Now, we need to create a vector to contain the target harvest acres (14000) over all periods, as we did for the row constraints in Section 9.3.6.1,

```
> target.acres <- rep(14000, 6)
```

This vector, of length T, represents the right-hand side (RHS) values giving the area, in acres, to be harvested in each period to achieve a strict area control forest.

Finally, use the `lpx_add_rows`,

```
> lpx_add_rows(leusch.lp, length(target.acres))
```

[1] 9

to add the *T* rows to the `leusch.lp` problem object. The function returns
the first index of the new row set.

Next, add the strict area control constraints, as was done in Section 9.3.6.1,

```
> val <- rep(1,8) ## a vector of 8 ones.
> for(i in 1:length(target.acres)) {
+    idx <-
+       as.numeric(rownames(subset(leusch.ylds, period == i)))
+    lpx_set_row_name(leusch.lp,
+                     i + length(stand.acres),
+                     paste("tac", i, sep="" ))
+    lpx_set_row_bnds(leusch.lp,
+                     i + length(stand.acres),
+                     LPX_FX,
+                     target.acres[i],
+                     target.acres[i])
+    lpx_set_mat_row(leusch.lp,
+                    i + length(stand.acres),
+                    length(val),
+                    idx,
+                    val)
+ }
```

We should now have added a total of 14 rows and 14 columns to the
problem object. Recall from Section 9.3.6.1 that we added only eight rows.
Here we added an additional six rows to represent the constraints on the total
acres cut (`tac`) in each of the six periods.

Again, using the `lpx_get_num_rows` and `lpx_get_num_cols` functions to
retrieve the number of rows and columns in the problem object,

```
> lpx_get_num_rows(leusch.lp)
```

[1] 14

```
> lpx_get_num_cols(leusch.lp)
```

[1] 48

This again verifies our results, and we now have a completed glpk problem
object that contains all the components of our forest planning problem.

9.3.7 Solving

function
Now we can solve the linear program by calling the `lpx_simplex` function:

```
> lpx_simplex(leusch.lp)

      0:    objval =    0.0000000e+00    infeas =    1.0000e+00 (0)
     13:    objval =    1.1584000e+06    infeas =    0.0000e+00 (1)
*    13:    objval =    1.1584000e+06    infeas =    0.0000e+00 (1)
*    25:    objval =    1.2246000e+06    infeas =    0.0000e+00 (1)
OPTIMAL SOLUTION FOUND
[1] 200
```

The `lpx_simplex` function is the wrapper function for the GLPK API `glp_simplex` routine, a driver to the glpk solver that uses the simplex method. This function retrieves problem data from the specified problem object, calls the glpk solver to solve the problem instance, and stores results of computations back into the problem object. Fortunately, this solution matches the solution reported by Leuschner (1990).

9.3.8 Results

In this section, we 1) obtain the value of the objective function, 2) decode the decision variables (column activities), 3) confirm that our solution achieves a strict area policy (equation (9.5)), 4) compute and plot the resulting wood-flow, and 5) examine shadow prices, reduced costs, and binding constraints. To conserve space, we include two functions that can be used to obtain row (`get.row.report`) and column (`get.col.report`) information from the linear programming object. See the provided script `schedpak.r` for these functions.

9.3.8.1 Objective Function

To extract the value of the objective function, use the `get_obj_val` function,

```
> leusch.obj <- lpx_get_obj_val(leusch.lp)
> leusch.obj
```

```
[1] 1224600
```

The resulting object, `leusch.obj`, is the value of equation 9.3 at its maximum. Therefore the optimal set of harvest decisions yields a total volume of $1224600 \times 10^3 \text{ft}^3$.

It would also be nice to have some report on the quality of the solution here as well (e.g., the Karush-Kuhn-Tucker (KKT) conditions (Karush, 1939; Kuhn and Tucker, 1951; Press et al., 2007)). The KKT general optimality conditions are necessary and sufficient conditions for the decision variables (**x**) to be optimal and are often used to report the quality of the solution for mathematical programming problems. However, at the time of this writing, the glpk package cannot produce them directly with R, but they can be generated and reported by the stand-alone solver glpsol if needed.

9.3.8.2 Decision Variables

To obtain the set of decision variables, we use the get.col.report function supplied in the FAR package,

```
> leusch.col.rpt <- get.col.report(leusch.lp)
```

where the leusch.col.rpt$activity contains the number of acres harvested in each stand for each period.

The get.col.report function provides the basic information on the columns in the solution. The function returns a data frame object that contains the output from the lpx_ wrapper functions lpx_get_col_name, lpx_get_col_stat, lpx_get_col_prim, lpx_get_col_lb, lpx_get_col_ub, lpx_get_col_dual, lpx_get_col_type, which return the column name, status, primary activity, lower bounds, upper bounds, dual value, and column type for each column in the problem. The get.row.report returns similar information for each row.

9.3.8.3 Harvested Area

To extract the harvest activity, we create a $T \times A$ matrix of the column activities. This object contains the area harvested from each stand over all periods,

```
> ac.per <-
+     matrix(as.numeric(as.matrix(I(leusch.col.rpt$activity))),
+            6, 8)
> ac.per
```

	[,1]	[,2]	[,3]	[,4]	[,5]	[,6]	[,7]	[,8]
[1,]	0	0	0	0	0	0	12000	2000
[2,]	0	0	0	0	0	14000	0	0
[3,]	0	0	0	0	8000	6000	0	0
[4,]	0	0	0	0	14000	0	0	0
[5,]	0	0	1000	5000	8000	0	0	0
[6,]	5000	5000	4000	0	0	0	0	0

where the rows represent the periods and the columns represent the stands. For example, $\mathscr{A}_5^x = (0,0,8000,14000,8000,0)$, meaning that stand five is to have 8000, 14000, and 8000 acres harvested in periods three, four, and five, respectively.

To check the solution, we can now determine the total acres harvested in each period obtained by summing the rows of ac.per,

```
> rowSums(ac.per)
```

```
[1] 14000 14000 14000 14000 14000 14000
```

which, upon inspection, matches our policy objectives, as defined in equation (9.5).

9.3.8.4 Woodflow

Next, compute the woodflow in each period, which can be calculated by hand. Computing the total volume in each period requires that we sum the products of the acres harvested by the yields per acre in each period.

The resulting vector is

```
> vol.cut.a <- c(199.4,197.4,203.4,211.4,217.6,195.4)
```

Then, plot the woodflow along with the results reported by Leuschner (1990) for even-flow constraints (Figure 9.1). The plot demonstrates that although an equal area is harvested over the planning horizon (equation (9.5)), the rate of volume extraction is not even over the planning horizon, which is in contrast to the strict volume control.

9.3.8.5 Reduced Costs

The *reduced cost* of a decision variable is the amount by which the objective function must change before that decision variable will enter the optimal solution (Leuschner, 1990). It provides information about the decision variables with a value of zero (Davis et al., 2001) and represents the opportunity cost of forcing a unit (e.g., an acre of $\mathscr{A}_{b=1}$) of the decision variable (e.g., a11) into the solution.

To obtain the reduced costs for the first period that stand 1, denoted $\mathscr{A}_{b=1}$, can be cut, we use the get.col.report function and return the first row,

```
> leusch.col.rpt[1,]
```

```
  name status activity lb ub dual b_ind type
1 a11       141        0  0  0 -4.8     0  111
```

which means that the yield per acre in $\mathscr{A}_{b=1}$ must increase from 3.2 to 8.0 hundred cubic feet ($8.0 = 3.2 + 4.8$), in order for it to be harvested in cutting period 1 (Leuschner, 1990).

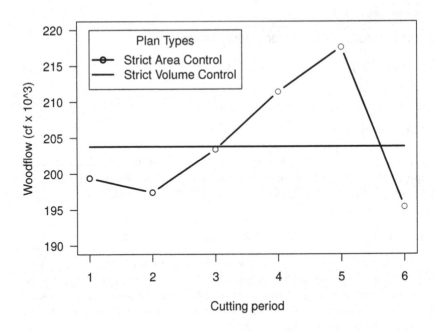

Fig. 9.1: Woodflow from the strict area control problem (Leuschner, 1990), where the strict area control represents the volume harvested given that a fixed proportion of the area is harvested each period. The flat line represents the even-flow solution, for which the harvest volume is 203.79×10^3 ft^3, presented in Leuschner (1990).

9.3.8.6 Slack/Surplus

The slack value is the amount on the right-hand side for a row constraint that is *not* used in the optimal solution, with a zero value indicating that the row is limiting or binding the solution (Davis et al., 2001). When this value is non-zero, a surplus of the constraint exists and the constraint does not influence the objective function value.

To obtain the slack and surplus values, use the `get.row.report`, which as does the `get.col.report` function in Section 9.3.8.2, returns a data frame that contains slack, surplus, and dual values (see Section 9.3.8.7), represented by the `lb`, `ub`, and `dual` columns of the data frame.

```
> leusch.row.rpt <- get.row.report(leusch.lp)
> leusch.row.rpt

  name status prim   lb   ub dual                    strerr
1   s1    142 5000    0 5000 12.9 variable with upper bound
```

2	s2	142	5000	0	5000	14.1	variable with upper bound
3	s3	142	5000	0	5000	15.1	variable with upper bound
4	s4	142	5000	0	5000	16.1	variable with upper bound
5	s5	142	30000	0	30000	17	variable with upper bound
6	s6	142	20000	0	20000	18	variable with upper bound
7	s7	142	12000	0	12000	19	variable with upper bound
8	s8	142	2000	0	2000	20	variable with upper bound
9	tac1	144	14000	14000	14000	-4.9	fixed variable
10	tac2	144	14000	14000	14000	-3.9	fixed variable
11	tac3	144	14000	14000	14000	-2.9	fixed variable
12	tac4	144	14000	14000	14000	-1.9	fixed variable
13	tac5	144	14000	14000	14000	-1	fixed variable
14	tac6	140	14000	14000	14000	0	fixed variable

From the printout, it appears that all stands have a zero upper bound for
the slack/surplus entries for the rows that define total possible areas that
can be harvested in each stand. Recall that in this case we used an equality
constraint so that *all* of the land was harvested (that is, a strict area control
schedule), so all of the slack variables for stand acres should be zero.

9.3.8.7 Shadow Prices

The shadow, or *dual*, price is the maximum amount the objective function
would change if one more unit of a constraint were included (Davis et al., 2001,
Chapter 6). In this case, the shadow price would represent our willingness to
pay for the ability to harvest an additional acre of land.

For example, there are physical constraints that limit the amount of land
available to the current landscape (84000 acres total, eight stands). What if
another acre of $\mathscr{A}_1$, or something very close to it, were available for harvest?
The shadow price would tell us how much an additional acre of land would
need to be worth in order to increase the objective function by including that
one additional acre.

To obtain the shadow prices, examine the column labeled *dual* in the output of the get.row.report function. We print the first row, which represents
the total land in $\mathscr{A}_1$,

```
> leusch.row.rpt[1,]
```

```
  name status prim lb   ub dual                         strerr
1   s1     142 5000  0 5000 12.9 variable with upper bound
```

The resulting value for leusch.row.rpt$dual is 12.9, which means that a
one acre increase in $\mathscr{A}_1$ harvested in the first period will increase the objective
function by $12.9 \times 10^3 \text{ft}^3$.

9.3.9 Archiving Problems

We have now formulated, generated, and solved the strict area schedule problem, and we now need to create a machine-readable archive of our model and solution. The glpk package has functions that can be used to archive problems, solve previously archived problems using glpk, and translate archived problems between solver formats like CPLEX (CPLEX Optimization Inc., 2003) and MathProg, which is a subset of AMPL (Fourer et al., 2003).

   To archive the leuschner problem object, save the problem object to your hard drive as an MPS file

```
> lpx_write_mps(leusch.lp, "leuschner.mps")
```

```
lpx_write_mps: writing problem data to `leuschner.mps'...
[1] 0
```

   function
or as a file that can be read by CPLEX (Figure 9.2)

```
> lpx_write_cpxlp(leusch.lp, "leuschner.xlp")
```

```
lpx_write_cpxlp: writing problem data to `leuschner.xlp'...
[1] 0
```

   function
which matches the example presented by Leuschner (1990, see Table 4.3).

   The resulting files (leuschner.mps and leuschner.xlp) can then be imported and solved from many solvers. For example, to solve and generate a machine-readable solution file for the same problem using the command-line solver glpsol, use

```
$ glpsol --cpxlp leuschner.xlp -w leuschner.sln
```

9.3.10 Cleanup

Our final step is to make sure that we free up the memory allocated for the problem by using the lpx_delete_prob function

```
> lpx_delete_prob(leusch.lp)
> leusch.lp
```

```
<pointer: 0x0>
```

which deletes the data associated with the problem object and sets an internal pointer to null (0x0).

```
\* Problem: leuschner *\

Maximize
 obj: + 3.2 a11 + 6.1 a12 + 8.3 a13 + 10.1 a14 + 11.6 a15 + 12.9 a16
 + 6.1 a21 + 8.3 a22 + 10.1 a23 + 11.6 a24 + 12.9 a25 + 14.1 a26
 + 8.3 a31 + 10.1 a32 + 11.6 a33 + 12.9 a34 + 14.1 a35 + 15.1 a36
 + 10.1 a41 + 11.6 a42 + 12.6 a43 + 14.1 a44 + 15.1 a45 + 16 a46
 + 11.6 a51 + 12.6 a52 + 14.1 a53 + 15.1 a54 + 16 a55 + 16.9 a56
 + 12.9 a61 + 14.1 a62 + 15.1 a63 + 16 a64 + 16.9 a65 + 17.7 a66
 + 14.1 a71 + 15.1 a72 + 16 a73 + 16.9 a74 + 17.7 a75 + 18.4 a76
 + 15.1 a81 + 16 a82 + 16.9 a83 + 17.7 a84 + 18.4 a85 + 19.1 a86

Subject To
 s1: + a16 + a15 + a14 + a13 + a12 + a11 <= 5000
 s2: + a26 + a25 + a24 + a23 + a22 + a21 <= 5000
 s3: + a36 + a35 + a34 + a33 + a32 + a31 <= 5000
 s4: + a46 + a45 + a44 + a43 + a42 + a41 <= 5000
 s5: + a56 + a55 + a54 + a53 + a52 + a51 <= 30000
 s6: + a66 + a65 + a64 + a63 + a62 + a61 <= 20000
 s7: + a76 + a75 + a74 + a73 + a72 + a71 <= 12000
 s8: + a86 + a85 + a84 + a83 + a82 + a81 <= 2000
 tac1: + a81 + a71 + a61 + a51 + a41 + a31 + a21 + a11 = 14000
 tac2: + a82 + a72 + a62 + a52 + a42 + a32 + a22 + a12 = 14000
 tac3: + a83 + a73 + a63 + a53 + a43 + a33 + a23 + a13 = 14000
 tac4: + a84 + a74 + a64 + a54 + a44 + a34 + a24 + a14 = 14000
 tac5: + a85 + a75 + a65 + a55 + a45 + a35 + a25 + a15 = 14000
 tac6: + a86 + a76 + a66 + a56 + a46 + a36 + a26 + a16 = 14000

End
```

Fig. 9.2: Archive for the problem presented by Leuschner (1990) in CPLEX format.

9.4 Summary

In this chapter, we presented and solved a simple linear programming forest estate planning model problem. For the problem, we presented procedures to 1) load the glpk package, 2) create the glpk problem object, 3) add decision variables (*columns*), 4) add constraints (*rows*), 5) solve the problem, 6) extract details about the solution directly, and finally 7) generate a machine-readable archive to create a complete digital record. These tasks are part of any estate planning problem and are commonly much more complex than the example that we have presented here. The techniques that we have described can be used to examine many forest estate planning and optimization problems, and readers are encouraged to apply them to their own data.

References

Agho, K., Dai, W., Robinson, J., 2005. Empirical saddlepoint approximations of the Studentized ratio and regression estimates for finite populations. Statistics and Probability Letters 71 (3), 237–247.

Armstrong, M., 1998. Basic Linear Geostatistics. Springer-Verlag, Berlin.

Avery, T. E., Burkhart, H. E., 2003. Forest Measurements, 5th Ed. McGraw–Hill, New York.

Bailey, R. L., Dell, T. R., 1973. Quantifying diameter distributions with the Weibull function. Forest Science 19, 97–104.

Banyard, S. G., 1987. Point sampling using constant tallies is biased: a tropical rainforest case study. Commonwealth Forestry Review 66, 161–163.

Bates, D. M., Watts, D. G., 1988. Nonlinear Regression Analysis and Its Applications. Wiley, New York.

Bell, J. F., Dillworth, J. R., 1997. Log Scaling and Timber Cruising. John Bell and Associates, Inc., P.O. Box 1538, Corvallis, Oregon 97339, 444 pp.

Berry, A. C., 1941. The accuracy of the Gaussian approximation to the sum of independent variates. Transactions of the American Mathematical Society 49 (1), 122–136.

Bettinger, P., Boston, K., Siry, J. P., Grebner, D. L., 2009. Forest Management and Planning. Elsevier, New York.

Bishop, C. M., 2006. Pattern Recognition and Machine Learning (Information Science and Statistics). Springer-Verlag, New York.

Borders, B., Souter, R., Bailey, R., Ware, K., 1987. Percentile-based distributions characterize forest stand tables. Forest Science 33 (2), 570–576.

Brackett, M. H., 2000. Data Resource Quality: Turning Bad Habits into Good Practices. Addison-Wesley information technology series. Addison-Wesley, Reading, MA.

Brazzale, A. R., 2005. hoa: An R package bundle for higher order likelihood inference. R News 5 (1).

Breiman, L., 2001a. Random forests. Machine Learning 45 (1), 5–32.

Breiman, L., 2001b. Statistical modeling: the two cultures. Statistical Science 16 (3), 199–231.

Briggs, D., 1994. Forest Products Measures and Conversion Factors: With Special Emphasis on the U.S. Pacific Northwest. Tech. Rep. Contribution No. 75, College of Forest Resources, University of Washington, Institute of Forest Resources.

Brillinger, D. R., 1964. The asymptotic behaviour of Tukey's general method of setting approximate confidence limits (the jackknife) when applied to maximum likelihood estimates. Revue de l'Institut International de Statistique 32 (3), 202–206.

Bruce, D., 1981. Consistent height-growth and growth rate estimates for re-measured plots. Forest Science 27, 711–725.

Buttrey, S., 2008. knncat: Nearest-neighbor classification with categorical variables. R package version 1.1.11.

Canty, A., Ripley, B., 2010. boot: Bootstrap R (S-Plus) functions. R package version 1.2-42.

Carr, D., Lewin-Koh, N., Maechler, M., 2010. hexbin: Hexagonal binning routines. R package version 1.22.0.

Casella, G., Berger, R. L., 1990. Statistical Inference. Duxbury Press, Belmont, CA.

Chambers, C. J., 1980. Empirical Growth and Yield Tables for the Douglas-fir Zone. Report 41, Department of Natural Resources, Olympia, WA.

Chambers, J. M., 1991. Data for models. In: Chambers, J. M., Hastie, T. J. (Eds.), Statistical Models in S. CRC Press, Inc., Boca Raton, FL.

Chambers, J. M., Hastie, T. J., 1992. Statistical Models in S. Chapman and Hall, London.

Christensen, R., Pearson, L. M., Johnson, W., 1992. Case-deletion diagnostics for mixed models. Technometrics 34 (1), 38–45.

Cleveland, W. S., 1993. Visualizing Data. Hobart Press, Lafayette, IN.

Clutter, J. L., Fortson, J. C., Pienaar, L. V., Brister, G. H., Bailey, R. L., 1983. Timber Management: A Quantitative Approach. Krieger Publishing Company, Malabar, FL.

Cochran, W. G., 1977. Sampling techniques. John Wiley and Sons, New York.

Commoner, B., 1971. The Closing Circle: Nature, Man, and Technology. Knopf, New York.

Cook, R. D., 1977. Detection of influential observations in linear regression. Technometrics 19, 15–18.

CPLEX Optimization Inc., 2003. CPLEX Linear Optimizer and Mixed Integer Optimizer. Incline Village, NV.

Cressie, N. A. C., 1993. Statistics for Spatial Data, Revised Edition. John Wiley and Sons, Inc., New York.

Curtis, R. O., 1982. A simple index for stand density. Forest Science 28 (1), 92–94.

Curtis, R. O., Clendenen, G. W., Reukema, D. L., DeMars, D. J., 1982. Yield tables for managed stands of coast Douglas-fir. General Technical Report PNW-135, Pacific Northwest Forest and Range Experiment Station, Forest Service, U.S. Department of Agriculture, Portland, Oregon.

Dantzig, G. B., 1963. Linear Programming and Extensions. Princeton University Press, Princeton, NJ.

Daubenmire, R., 1952. Forest vegetation of Northern Idaho and adjacent Washington, and its bearing on concepts of vegetation classification. Ecological Monographs 22, 301–330.

David, H. A., 1995. First (?) occurrence of common terms in mathematical statistics. American Statistician 49 (2), 121–133.

Davis, L. S., Johnson, K. N., Bettinger, P. S., Howard, T. E., 2001. Forest Management, 4th Edition. McGraw-Hill Series in Forest Resources. McGraw-Hill, New York.

Davison, A. C., Hinkley, D. V., 1997. Bootstrap Methods and Their Application. Cambridge University Press, Cambridge.

de Gruijter, J. J., Brus, D. J., Bierkins, M. F. P., Knotters, M., 2006. Sampling for Natural Resource Monitoring. Springer-Verlag.

Demidenko, E., 2004. Mixed Models: Theory and Applications. John Wiley and Sons, Hoboken, NJ.

Demidenko, E., Stukel, T. A., 2005. Influence analysis for linear mixed-effects models. Statistics in Medicine 24, 893–909.

Dempster, A. P., Laird, N. M., Rubin, D. B., 1977. Maximum likelihood from incomplete data via the EM algorithm (with discussion). Journal of the Royal Statistical Society (B) 39 (1), 1–38.

Draper, N. R., Smith, H., 1998. Applied Regression Analysis, 3rd Edition. Jonh Wiley and Sons, New York.

Drew, T. J., Flewelling, J. W., 1979. Stand density management: an alternative approach and its application to Douglas-fir plantations. Forest Science 25 (3), 518–532.

DuPont, 2005. DuPont Oust® Extra. Material Safety Data Sheet, http://msds.dupont.com/msds/pdfs/EN/PEN_09004a2f8061d37c.pdf, Accessed February 8, 2006.

Duursma, R., Marshall, J. D., Robinson, A. P., 2003. Leaf area index inferred from solar beam transmission in mixed conifer forests on complex terrain. Agricultural and Forest Meteorology 118, 221–236.

Efron, B., 1979. Bootstrap methods: Another look at the jackknife. The Annals of Statistics 7 (1), 1–26.

Efron, B., Tibshirani, R. J., 1993. An Introduction to the Bootstrap. No. 57 in Monographs on Statistics and Applied Probability. Chapman and Hall, New York.

Esseen, C.-G., 1942. On the Liapunoff limit of error in the theory of probability. Arkiv för matematik, astronomi och fysik A28, 1–19.

Fitzmaurice, G., Laird, N., Ware, J., 2004. Applied Longitudinal Analysis. Wiley–Interscience, New York.

Flewelling, J. W., Pienaar, L. V., 1981. Multiplicative Regression with Lognormal Errors. Forest Science 27 (2), 281–289.

Fourer, R., Gay, D. M., Kernighan, B. W., 2003. AMPL: A modeling language for mathematical programming, 2nd Edition. Thomson,

Toronto, Canada, first chapter, software, and other material available at http://www.ampl.com.

Froese, R. E., 2003. Re-engineering the Prognosis basal area increment model for the Inland Empire. Ph.D. thesis, University of Idaho.

Furrer, R., Nychka, D., Sain, S., 2009. fields: Tools for spatial data. R package version 6.01.

Gallant, A. R., 1987. Nonlinear Statistical Models. John Wiley & Sons, New York.

García, O., 1983. A stochastic differential equation model for the height growth of forest stands. Biometrics 39 (4), 1059–1072.

García, O., 1992. What is a diameter distribution? In: Minowa, M., Tsuyuki, S. (Eds.), Proceedings of the Symposium on Integrated Forest Management Information Systems. Japan Society of Forest Planning Press, Tokyo, pp. 11–29.

Gelman, A., 2005. Analysis of variance — why it is more important than ever. The Annals of Statistics 33, 1–31.

Gelman, A., Hill, J., 2007. Data Analysis Using Regression and Multi-level/Hierarchical Models. Cambridge University Press, Cambridge.

Goodman, L. A., 1960. On the exact variance of products. Journal of the American Statistical Association 55, 708–713.

Goovaerts, P., 1997. Geostatistics for Natural Resources Evaluation. Applied Geostatistics Series. Oxford University Press, New York, New York.

Goutis, C., Casella, G., 1999. Explaining the saddlepoint approximation. American Statistician 53 (3), 216–224.

Gove, J. H., 2003. Moment and maximum likelihood estimators for Weibull distributions under length- and area-biased sampling. Environmental and Ecological Statistics 10, 455–467.

Haara, A., Maltamo, M., Tokola, T., 1997. K-nearest-neighbor method for estimating basal-area diameter distribution. Scandinavian Journal of Forest Research 12, 200–208.

Hafley, W. L., Schreuder, H. T., 1977. Statistical distributions for fitting diameter and height data in even-aged stands. Canadian Journal of Forest Research 7, 481–487.

Hall, P., 1992. The Bootstrap and Edgeworth Expansion. Springer-Verlag, New York.

Hamann, J. D., Ritchie, M. W., 2009. rconifers: R interface to the CONIFERS forest growth model. R package version 1.0.0.

Hann, D. W., Wang, C. H., 1990. Mortality equations for individual trees in southwest Oregon. Tech. Rep. Research Bulletin 67, Oregon State University, Forest Research Laboratory, Corvallis, OR.

Hanus, M. L., Marshall, D. D., Hann, D. W., 1999. Height–diameter equations for six species in the coastal regions of the Pacific Northwest. Research Contribution 25, Forest Research Laboratory, Oregon State University, Corvallis, OR.

Harrell, F. E., 2001. Regression Modeling Strategies: With Applications to Linear Models, Logistic Regression and Survival Analysis. Springer, New York.

Hastie, T., Tibshirani, R., Friedman, J., 2009. The Elements of Statistical Learning: Data Mining, Inference, and Prediction, 2nd Edition. Springer, New York.

Holmström, H., 2002. Estimation of single-tree characteristics using the knn method and plotwise aerial photograph interpretations. Forest Ecology and Management 167, 303–314.

Holmström, H., Kallur, H., Støahl, G., 2003. Cost-plus-loss analyses of forest inventory strategies based on kNN-assigned reference sample plot data. Silva Fennica 37 (3), 381–398.

Horvitz, D., Thompson, D., 1952. A generalization of sampling without replacement from a finite population. Journal of the American Statistical Association 47, 89–96.

Husch, B., Beers, T., Kershaw Jr., J., 2003. Forest Mensuration. John Wiley and Sons, Hoboken, NJ.

Ihaka, R., Gentleman, R., 1996. R: A language for data analysis and graphics. Journal of Computational and Graphical Statistics 5 (3), 299–314.

Iles, K., 2003. A Sampler of Inventory Topics. K. Iles and Associates, Inc., Nanaimo.

Isaaks, E. H., Srivastava, R. M., 1989. An Introduction to Applied Geostatistics. Oxford University Press, Inc., New York.

James, D. A., DebRoy, S., 2009. RMySQL: R interface to the MySQL database. R package version 0.7-4.

Jensen, J. L., 1906. Sur les fonctions convexes et les inégualités entre les valeurs moyennes. Acta Mathematica 30, 175–193 (in French).

Johnson, N. L., Kotz, S., Balakrishnan, N., 1994. Continuous univariate distributions, 2nd Edition. Wiley–Interscience, New York.

Jones, H. L., 1956. Investigating the properties of a sample mean by employing random subsample means. Journal of the American Statistical Association 51, 54–83.

Jones, L. V. (Ed.), 1986. Data analysis and behavioral science or learning to bear the quantitative man's burden by shunning badmandments. The Collected Works of John W. Tukey, Philosophy and Principles of Data Analysis 1949–1964, Volume III. Chapman and Hall/CRC, Boca Raton, FL., pp. 187–390.

Journel, A., Huijbregts, C., 1978. Mining Geostatistics. Academic Press, London.

Karush, W., 1939. Minima of functions of several variables with inequalities as side constraints. Master's thesis, Department of Mathematics, University of Chicago.

King, J. E., 1966. Site index curves for douglas-fir in the pacific northwest. Weyerhaeuser Forestry Paper 8, Weyerhaeuser Co. Centralia, WA.

Kline, K., Kline, D., 2001. SQL in a Nutshell. O'Reilly & Associates, Inc., Sebastapol, CA.

Korhonen, K., Kangas, A., 1997. Application of nearest-neighbor regression for generating sample tree information. Scandinavian Journal of Forest Research 12, 97–101.

Kozak, A., 1988. A variable-exponent taper equation. Canadian Journal of Forest Research 18, 1363–1368.

Kozak, A., Munro, D. D., Smith, J., 1968. Taper functions and thier application in forest inventory. Forestry Chronicle 45, 278–283.

Kuhn, H. W., Tucker, A. W., 1951. Nonlinear programming. In: Neyman, J. (Ed.), Proceedings of the Second Berkeley Symposium on Mathematical Statistics and Probability, Berkeley, 1950. University of California Press, Berkeley, CA., pp. 481–492, reprinted in: Readings in Mathematical Economics, Vol 1, Value Theory, (P. Newman, Ed.), The Johns Hopkins University Press, Baltimore, 1968, pp. 3–14.

Kuroda, M., 2004. Markov Chain Monte Carlo. In: Watanabe, M., Yamaguchi, K. (Eds.), The EM Algorithm and Related Statistical Models. Marcel Dekker, Inc., New York, Chapter 9.

Laird, N. M., Ware, J. H., 1982. Random-effects models for longitudinal data. Biometrics 38, 963–974.

Landsat Project Science Office, 2005. The Landsat-7 science data user's handbook [online]. Landsat Project Science Office, NASA Goddard Space Flight Center, Greenbelt, MD, http://ltpwww.gsfc.nasa.gov/IAS/handbook/handbook_toc.html. Accessed December 28, 2005.

Lee, L., Luangkesorn, L., 2010. glpk: GNU linear programming kit. R package version 4.8-0.5.

Lee, Y., Nelder, J. A., Pawitan, Y., 2006. Generalized Linear Models with Random Effects: Unified Analysis via H-likelihood. Chapman and Hall/CRC, Boca Raton, FL.

Leuschner, W. A., 1990. Forest Regulation, Harvest Scheduling, and Planning Techniques. John Wiley and Sons, Inc., New York.

Lewin-Koh, N. J., Bivand, R., 2010. maptools: Tools for reading and handling spatial objects. R package version 0.7-34.

Little, R. J. A., Rubin, D. B., 2002. Statistical Analysis with Missing Data. John Wiley and Sons, Inc., New York.

Longford, N. R., 1993. Random Coefficient Models. No. 11 in Oxford Statistical Science Series. Oxford University Press, Inc., New York.

Lumley, T., 2004. Analysis of complex survey samples. Journal of Statistical Software 9 (8), 1–19.

Lumley, T., 2010. survey: Analysis of complex survey samples. R package version 3.22-1.

Madras, N., 2002. Lectures on Monte Carlo Methods. No. 16 in Fields Institute Monographs. American Mathematical Society, Providence, RI.

Magnussen, S., 2001. Saddlepoint approximations for statistical inference of PPP sample estimates. Scandinavian Journal of Forest Research 16 (2), 180–192.

Makhorin, A., 2009. GNU linear programming kit, version 4.9. GNU Software Foundation, http://www.gnu.org/software/glpk/glpk.html.

Manly, B. F., 1997. Randomization, Bootstrap, and Monte Carlo Methods in Biology, 2nd Edition. Chapman and Hall, London.

Marti, K., 2005. Stochastic Optimization Methods. Springer, New York.

McArdle, R. E., Meyer, W. H., Bruce, D., 1949. The yield of Douglas-fir in the Pacific Northwest. Technical Bulletin No. 201, United States Department of Agriculture, Washington, D.C.

McLachlan, G. J., Krishnan, T., 2008. The EM Algorithm and Extensions. John Wiley & Sons, Inc., Hoboken, NJ.

Miller, Jr., R. G., 1964. A trustworthy jackknife. The Annals of Mathematical Statistics 35 (4), 1594–1605.

Moeur, M., Stage, A. R., 1995. Most similar neighbor: An improved sampling inference procedure for natural resource planning. Forest Science 41 (2), 337–359.

Nelder, J. A., Mead, R., 1965. A simplex algorithm for function minimization. Computer Journal 7, 308–313.

Neteler, M., Mitasova, H., 2002. Open Source GIS: A GRASS GIS approach. Kluwer Academic Publishers, Dordrecht.

Newton, M., Hanson, T. J., 1998. Bias in site estimation from early competition. In: 19th Forest Vegetation Management Conference. Redding, CA., pp. 78–84.

Nocedal, J., Wright, S. J., 2006. Numerical Optimization, 2nd Edition. Springer-Verlag, New York.

Novo, A. A., Schafer, J. L., 2002. norm: Analysis of multivariate normal datasets with missing values. R package version 1.0-9.

Ohmann, J. L., Gregory, M. J., 2002. Predictive mapping of forest composition and structure with direct gradient analysis and nearest-neighbor imputation in coastal Oregon, USA. Canadian Journal of Forest Research 32 (4), 725–741.

Okabe, A., Boots, B., Sugihara, K., Chiu, S. N., 2000. Spatial Tessellations: Concepts and Applications of Voronoi Diagrams, 2nd Edition. John Wiley and Sons, Inc., New York.

Oksanen, J., Blanchet, F. G., Kindt, R., Legendre, P., O'Hara, R. B., Simpson, G. L., Solymos, P., Stevens, M. H. H., Wagner, H., 2010. vegan: Community Ecology Package. R package version 1.17-2.

Pawitan, Y., 2001. In All Likelihood: Statistical Modelling and Inference Using Likelihood. Clarendon Press, Oxford.

Pebesma, E. J., 2004. Multivariable geostatistics in S: the gstat package. Computers & Geosciences 30 (7), 683–691.

Pinheiro, J. C., Bates, D. M., 2000. Mixed-effects models in S and Splus. Springer-Verlag, New York.

Pocewicz, A. L., Gessler, P., Robinson, A. P., 2004. The relationship between effective plant area index and landsat spectral response across elevation, solar insolation, and spatial scales in a northern Idaho forest. Canadian Journal of Forest Research 34, 465–480.

Prayaga, S. K., Eddelbuettel, D., Tiffin, N., Conway, J., 2009. RPostgreSQL: R interface to the PostgreSQL database system. R package version 0.1-6.

Press, W. H., Flannery, B. P., Teukolsky, S. A., Vetterling, W. T., 2007. Numerical Recipes: The Art of Scientific Computing, 3rd Edition. Cambridge University Press, Cambridge.

Quenouille, M. H., 1949a. Approximate tests of correlation in time-series. Journal of the Royal Statistical Society Series B 11, 68–84.

Quenouille, M. H., 1949b. Problems in plane sampling. Annals of Mathematical Statistics 20, 355–375.

Quenouille, M. H., 1956. Notes on bias in estimation. Biometrika 43, 353–360.

R core members, DebRoy, S., Bivand, R., et al., 2010. foreign: Read Data Stored by Minitab, S, SAS, SPSS, Stata, Systat, dBase, ... R package version 0.8-40.

R Development Core Team, 2010. R: A Language and Environment for Statistical Computing. R Foundation for Statistical Computing, Vienna.

Ratkowsky, D. A., 1983. Nonlinear Regression Modeling: A Unified Practical Approach. Vol. 48 of Statistics, Textbooks and Monographs. Marcel Dekker, Inc., New York.

Reid, N., 1988. Saddlepoint methods and statistical inference. Statistical Science 3, 213–238.

Reid, N., 2003. Asymptotics and the theory of inference. Annals of Statistics 31 (6), 1695–1731.

Reid, N., Fraser, D. A. S., 2000. Higher-order asymptotics: costs and benefits. In: Rao, C. R., Székely, G. J. (Eds.), Statistics for the 21st Century: Methodologies for Applications of the Future. Marcel Dekker, Inc., New York, pp. 351–365.

Reineke, L. H., 1933. Perfecting a stand-density index for even-aged forests. Journal of Agricultural Research 46 (7), 627–638.

Ripley, B., Lapsley, M., 2009. RODBC: ODBC database access. R package version 1.3-1.

Ritchie, M. W., Hamann, J. D., 2006. Modeling dynamics of competing vegetation in young conifer plantations of northern California and southern Oregon, USA. Canadian Journal of Forest Research 36 (10), 2523–2532.

Ritchie, M. W., Hamann, J. D., 2008. Individual-tree height-, diameter- and crown-width increment equations for young Douglas-fir plantations. New Forests 35 (2), 173–186.

Robinson, G. K., 1991. That BLUP is a good thing: the estimation of random effects. Statistical Science 6 (1), 15–32.

Rönnqvist, M., 2003. Optimization in forestry. Mathematical Programming 97 (1), 267–284.

Sarkar, D., 2010. lattice: Lattice Graphics. R package version 0.18-8.

Särndal, C.-E., Swensson, B., Wretman, J., 1992. Model Assisted Survey Sampling. Springer-Verlag, New York, 694 p.

Schabenberger, O., 2005. Mixed model influence diagnostics. In: SUGI 29. Paper 189-29, SAS Institute.

Schabenberger, O., Gotway, C., 2005. Statistical Methods for Spatial Data Analysis. Chapman & Hall/CRC, Boca Raton, 488 p.

Schabenberger, O., Pierce, F. J., 2002. Contemporary statistical models for the plant and soil sciences. CRC Press, Boca Raton, FL.

Schliep, K., Hechenbichler, K., 2009. kknn: Weighted k-nearest neighbors. R package version 1.0-7.

Schreuder, H. T., Gregoire, T. G., Wood, G. B., 1993. Sampling methods for multiresource forest inventory. John Wiley & Sons, Inc., New York.

Scott, D. W., 1985. Average shifted histograms: effective nonparametric density estimators in several dimensions. Annals of Statistics 13 (3), 1024–1040.

Seber, G., Wild, C., 2003. Nonlinear Regression. John Wiley & Sons, Inc., New York.

Shao, J., Tu, D., 1995. The Jacknife and Bootstrap. Springer, New York.

Stage, A. R., 1963. A mathematical approach to polymorphic site index curves for Grand fir. Forest Science 9 (2), 167–180.

Tukey, J., 1962. The future of data analysis. Annals of Mathematical Statistics 33, 1–67.

Tukey, J. W., 1958. Bias and confidence in not-quite large samples (abstract of preliminary report). Annals of Mathematical Statistics 29, 614.

Vanclay, J. K., 1994. Modelling Forest Growth and Yield: Applications to Mixed Tropical Forests. CAB International, Oxon.

Venables, W. N., Ripley, B. D., 2002. Modern Applied Statistics with S. 4th Edition. Springer, New York.

von Guttenberg, A. R., 1915. Growth and yield of spruce in Hochgebirge. Franz Deuticke, Vienna (in German).

Warnes, G. R., 2010. gmodels: Various R programming tools for model fitting. R package version 2.15.0.

Watanabe, M., Yamaguchi, K. (Eds.), 2004. The EM Algorithm and Related Statistical Models. Marcel Dekker, Inc., New York.

Webster, R., Oliver, M., 2001. Geostatistics for Environmental Scientists. John Wiley & Sons, Inc., Chichester.

Weintraub, A., Navon, D., 1976. A forest management planning model integrating silvicultural and transportation activities. Management Science 22 (12), 1299–1309.

Weisberg, S., 2005. Applied Linear Regression, 3rd Edition. Wiley-Interscience, New York, 310 p.

Wickham, H., 2009. ggplot2: Elegant Graphics for Data Analysis. Springer, New York.

Wolter, K. M., 1985. Introduction to Variance Estimation. Springer-Verlag, New York.

Wood, S. N., 2006. Generalized Additive Models: An Introduction with R. Chapman and Hall/CRC, Boca Raton, FL.

Wykoff, W. R., Crookston, N. L., Stage, A. R., 1982. User's Guide to the Stand Prognosis Model. GTR-INT 133, USDA Forest Service, Ogden, UT.

Zeide, B., 1993. Analysis of growth equations. Forest Science 39 (3), 549–616.

Index